T0339949

Analysis of Nanoplastics and Microplastics in Food

Food Analysis & Properties

Series Editor
Leo M. L. Nollet
University College Ghent, Belgium

This CRC series **Food Analysis and Properties** is designed to provide state-of-the-art coverage on topics to the understanding of physical, chemical and functional properties of foods, including: (1) recent analysis techniques of a choice of food components; (2) developments and evolutions in analysis techniques related to food; (3) recent trends in analysis techniques of specific food components and/or a group of related food components.

Flow Injection Analysis of Food Additives

Edited by Claudia Ruiz-Capillas and Leo M. L. Nollet

Marine Microorganisms: Extraction and Analysis of Bioactive Compounds
Edited by Leo M. L. Nollet

Multiresidue Methods for the Analysis of Pesticide Residues in Food
Edited by Horacio Heinzen, Leo M.L. Nollet, and Amadeo R. Fernandez-Alba

Spectroscopic Methods in Food Analysis
Edited by Adriana S. Franca and Leo M.L. Nollet

Phenolic Compounds in Food: Characterization and Analysis
Edited by Leo M.L. Nollet, Janet Alejandra Gutierrez-Uribe

Testing and Analysis of GMO-containing Foods and Feed
Edited by Salah E. O. Mahgoub, Leo M. L. Nollet

Fingerprinting Techniques in Food Authenticity and Traceability
Edited by K.S. Siddiqi and Leo M.L. Nollet

Hyperspectral Imaging Analysis and Applications for Food Quality
Edited by Nrusingha Charan Basantia, Leo M.L. Nollet, Mohammed Kamruzzaman

Ambient Mass Spectroscopy Techniques in Food and the Environment
Edited by Leo M.L. Nollet and Basil K. Munjanja

Food Aroma Evolution: During Food Processing, Cooking and Aging
Edited by Matteo Bordiga, Leo M. L. Nollet

Proteomics for Food Authentication
Edited by Leo M.L. Nollet, Semih Otles

Mass Spectrometry Imaging in Food Analysis
Edited by Leo M. L. Nollet

Analysis of Nanoplastics and Microplastics in Food
Edited by Leo M. L. Nollet and Khwaja Salahuddin Siddiqi

For more information, please visit the Series Page: https://www.crcpress.com/Food-Analysis--Properties/book-series/CRCFOODANPRO

Analysis of Nanoplastics and Microplastics in Food

Edited by
Leo M. L. Nollet
Khwaja Salahuddin Siddiqi

CRC Press
Taylor & Francis Group
Boca Raton London New York

CRC Press is an imprint of the
Taylor & Francis Group, an **informa** business

First edition published 2021
by CRC Press
6000 Broken Sound Parkway NW, Suite 300, Boca Raton, FL 33487-2742

and by CRC Press
2 Park Square, Milton Park, Abingdon, Oxon, OX14 4RN

© 2021 Taylor & Francis Group, LLC

CRC Press is an imprint of Taylor & Francis Group, LLC

Library of Congress Cataloging-in-Publication Data

Names: Nollet, Leo M. L., 1948- editor. | Siddiqi, K. S., editor.
Title: Analysis of nanoplastics and microplastics in food / edited by Leo M.L. Nollet, Khwaja Salahuddin Siddiqi.
Other titles: Food analysis and properties, 2475-7551
Description: First edition. | Boca Raton : CRC Press, 2020. | Series: Food analysis and properties | Includes bibliographical references and index.
Identifiers: LCCN 2020012390 (print) | LCCN 2020012391 (ebook) | ISBN 9781138600188 (hardback) | ISBN 9780429469596 (ebook)
Subjects: MESH: Microplastics--analysis | Food Contamination--analysis | Microplastics--adverse effects | Water Pollutants, Chemical--analysis | Environmental Monitoring--methods
Classification: LCC TD427.P62 (print) | LCC TD427.P62 (ebook) | NLM WA 701 | DDC 363.738--dc23
LC record available at https://lccn.loc.gov/2020012390
LC ebook record available at https://lccn.loc.gov/2020012391

ISBN: 978-1-138-60018-8 (hbk)
ISBN: 978-0-429-46959-6 (ebk)

Typeset in Sabon
by Nova Techset Private Limited, Bengaluru & Chennai, India

Contents

SECTION IV   SAMPLING AND ANALYSIS

SECTION V   MICROPLASTICS AND POPs

SECTION VI   MICROPLASTICS IN FRESHWATER

Series Preface

There will always be a need for analyzing methods of food compounds and properties. Current trends in analyzing methods include automation, increasing the speed of analyses and miniaturization. The unit of detection has evolved over the years from micrograms to picograms.

A classical pathway of analysis is sampling, sample preparation, cleanup, derivatization, separation and detection. At every step, researchers are working and developing new methodologies. A large number of papers are published every year on all facets of analysis. So, there is a need for books that gather information on one kind of analysis technique or on analysis methods of a specific group of food components.

The scope of the CRC Series on *Food Analysis & Properties* aims to present a range of books edited by distinguished scientists and researchers who have significant experience in scientific pursuits and critical analysis. This series is designed to provide state-of-the-art coverage on topics such as

- Recent analysis techniques on a range of food components
- Developments and evolution in analysis techniques related to food
- Recent trends in analysis techniques of specific food components and/or a group of related food components
- The understanding of physical, chemical and functional properties of foods

The book *Analysis of Nanoplastics and Microplastics in Food* is volume number 11 of this series.

I am happy to be a series editor of such books for the following reasons:

- I am able to pass on my experience in editing high-quality books related to food.
- I get to know colleagues from all over the world more personally.
- I continue to learn about interesting developments in food analysis.

A lot of work is involved in the preparation of a book. I have been assisted and supported by a number of people, all of whom I would like to thank. I would especially like to thank the team at CRC Press/Taylor & Francis Group, with a special word of thanks to Steve Zollo, senior editor.

Many, many thanks to all the editors and authors of this volume and future volumes. I very much appreciate all their effort, time and willingness to do a great job.

I dedicate this series to

- My wife, for her patience with me (and all the time I spend on my computer)

- All patients suffering from prostate cancer; knowing what this means, I am hoping they will have some relief

Dr. Leo M. L. Nollet (retired)
University College Ghent
Ghent, Belgium

Preface

The world's ever-increasing use of plastics has created large areas of floating plastic waste in the oceans so-called plastic soup. Areas as big as France have been observed. This floating plastic debris is gradually fragmenting into smaller particles that eventually become microplastics and even nanoplastics. These may have the forms of pellets, flakes, spheroids and beads.

Microplastics range in size from 0.1 to 5000 micrometers (μm), or 5 millimeters. Nanoplastics measure from 0.001 to 0.1 μm (i.e., 1 to 100 nanometers).

There are some data on nanoplastics and microplastics in food, particularly for the marine environment. Fish show high concentrations, but because microplastics are mostly present in the stomach and intestines, they are usually removed and consumers are not exposed to them. However, with crustaceans and bivalve mollusks like oysters and mussels, one can eat the digestive tract so there is potential for exposure. Microplastics have also been reported in honey, beer and table salt.

One concern is that concentrations of pollutants like polychlorinated biphenyls (PCBs), polycyclic aromatic hydrocarbons (PAHs) and other persistent organic pollutants (POPs) may accumulate in, to or from microplastics. There might also be residues of compounds, such as bisphenol A (BPA), used in packaging. Some studies suggest that after consuming microplastics in food, these substances may transfer into human tissues. Thus, it is important to estimate the average intake.

It is known that engineered nanoparticles (from different types of nanomaterials) can enter human cells, and this may have consequences for human health.

This book consists of six sections:

In Section 1, "Microplastics and Nanoplastics," general information on plastic residues is given. Chapter 2 deals with impacts and management of plastic waste in the aquatic environment.

In Section 2, "Plastics in the Environment," we have three chapters. Chapter 3 discusses the behavior of plastics in all compartments of the environment. In Chapter 4, we go deeper on the fate of plastics in the oceans. Chapter 5 provides information on plastics in the intertidal zone.

Chapter 6 of Section 3 details how plastics may end up in foods.

Many sampling methods and analysis methods of plastics from different media, and especially from foods, are developed. That information is compiled in Section 4, "Sampling and Analysis."

Marine litter plastics and microplastics may attract or originate toxic chemical components, persistent organic pollutants. This item is discussed in two chapters of Section 5.

Two chapters of the last section deal with plastics in a specific medium: freshwater.

The editors are very happy to thank all colleagues who contributed to this volume. Their efforts will be appreciated forever.

To finish this volume, a lot of time and patience was needed from the editors. But perseverance has one nice consequence: the result.

Leo M. L. Nollet
Khwaja Salahuddin Siddiqi

Great works are performed, not by strength, but by perseverance.

Samuel Johnson

Editors

Leo M. L. Nollet earned an MS (1973) and PhD (1978) in biology from the Katholieke Universiteit Leuven, Belgium. He is an editor and associate editor of numerous books. He edited for M. Dekker, New York—now CRC Press of Taylor & Francis Group—the first, second and third editions of *Food Analysis by HPLC* and *Handbook of Food Analysis*. The last edition is a two-volume book. Dr. Nollet also edited the *Handbook of Water Analysis* (first, second and third editions) and *Chromatographic Analysis of the Environment*, third and fourth editions (CRC Press). With F. Toldrá, he coedited two books published in 2006, 2007 and 2017: *Advanced Technologies for Meat Processing* (CRC Press) and *Advances in Food Diagnostics* (Blackwell Publishing—now Wiley). With M. Poschl, he coedited the book *Radionuclide Concentrations in Foods and the Environment*, also published in 2006 (CRC Press). Dr. Nollet also coedited, with Y. H. Hui and other colleagues, several books: *Handbook of Food Product Manufacturing* (Wiley, 2007), *Handbook of Food Science, Technology, and Engineering* (CRC Press, 2005), *Food Biochemistry and Food Processing* (first and second editions; Blackwell Publishing—now Wiley—2006 and 2012), and *Handbook of Fruits and Vegetable Flavors* (Wiley, 2010). In addition, he edited *Handbook of Meat, Poultry and Seafood Quality*, first and second editions (Blackwell Publishing—now Wiley—2007 and 2012). From 2008 to 2011, he published five volumes on animal product-related books with F. Toldrá: *Handbook of Muscle Foods Analysis*, *Handbook of Processed Meats and Poultry Analysis*, *Handbook of Seafood and Seafood Products Analysis*, *Handbook of Dairy Foods Analysis*, and *Handbook of Analysis of Edible Animal By-Products*. Also, in 2011, with F. Toldrá, he coedited two volumes for CRC Press: *Safety Analysis of Foods of Animal Origin* and *Sensory Analysis of Foods of Animal Origin*. In 2012, they published the *Handbook of Analysis of Active Compounds in Functional Foods*. In a coedition with Hamir Rathore, *Handbook of Pesticides: Methods of Pesticides Residues Analysis* was marketed in 2009; *Pesticides: Evaluation of Environmental Pollution* in 2012; *Biopesticides Handbook* in 2015; and *Green Pesticides Handbook: Essential Oils for Pest Control* in 2017. Other finished book projects include *Food Allergens: Analysis, Instrumentation, and Methods* (with A. van Hengel; CRC Press, 2011) and *Analysis of Endocrine Compounds in Food* (Wiley-Blackwell, 2011). Dr. Nollet's recent projects include *Proteomics in Foods* with F. Toldrá (Springer, 2013) and *Transformation Products of Emerging Contaminants*

in the Environment: Analysis, Processes, Occurrence, Effects and Risks with D. Lambropoulou (Wiley, 2014). In the series *Food Analysis & Properties*, he edited (with C. Ruiz-Capillas) *Flow Injection Analysis of Food Additives* (CRC Press, 2015) and *Marine Microorganisms: Extraction and Analysis of Bioactive Compounds* (CRC Press, 2016). With A. S. Franca, he coedited *Spectroscopic Methods in Food Analysis* (CRC Press, 2017), and with Horacio Heinzen and Amadeo R. Fernandez-Alba, he coedited *Multiresidue Methods for the Analysis of Pesticide Residues in Food* (CRC Press, 2017). Further volumes in the series *Food Analysis & Properties* are *Phenolic Compounds in Food: Characterization and Analysis* (with Janet Alejandra Gutierrez-Uribe, 2018), *Testing and Analysis of GMO-Containing Foods and Feed* (with Salah E. O. Mahgoub, 2018), *Fingerprinting Techniques in Food Authentication and Traceability* (with K. S. Siddiqi, 2018), *Hyperspectral Imaging Analysis and Applications for Food Quality* (with N. C. Basantia and Mohammed Kamruzzaman, 2018), *Ambient Mass Spectroscopy Techniques in Food and the Environment* (with Basil K. Munjanja, 2019), and *Food Aroma Evolution: During Food Processing, Cooking, and Aging* (with M. Bordiga, 2019).

Professor **Khwaja Salahuddin Siddiqi** is former chairman and professor emeritus of the Department of Chemistry, Aligarh Muslim University, Aligarh, India (PhD 1973, MPhil 1971, MSc 1969; Aligarh Muslim University, Aligarh, UP, India). His specialization is inorganic chemistry. Presently, he is an honorary professor, Faculty of Unani Medicine, Aligarh Muslim University, Aligarh, India. His research interests are transition metal chemistry, coordination chemistry, organometallic chemistry, bioinorganic chemistry and nanochemistry. He is the author of numerous papers and book chapters. He is a member of diverse societies and associations.

Contributors

A. P. Bangun
Department of Aquatic Resources
  Management
Faculty of Agriculture
Universitas Sumatera Utara
Medan, North Sumatera, Indonesia

D. M. Bila
Departamento de Engenharia Sanitária
  e do Meio Ambiente da Faculdade de
  Engenharia
Universidade do Estado do Rio de Janeiro
Rio de Janeiro, Brasil

M. Cole
College of Life and Environmental
  Sciences: Bioscience
University of Exeter
Exeter, United Kingdom

Gift Samuel David
Department of Fisheries and Aquatic
  Environmental Management
University of Uyo
Uyo, Akwa Ibom State, Nigeria

Vividha Dhapte-Pawar
Department of Pharmaceutics
Poona College of Pharmacy
Bharati Vidyapeeth University
Maharashtra, India

M. Humam Zaim Faruqi
Department of Chemical Engineering
Faculty of Engineering
  and Technology
Aligarh Muslim University
Aligarh, India

Cristina Fossi
Department of Environmental Sciences
University of Siena
Siena, Italy

Frederic Gallo
SCP/RAC, Barcelona Convention for
  the Protection of the Marine
  Environment and the Coastal
  Region of the Mediterranean
Stockholm Convention Regional Activity
  Centre in Spain
Barcelona, Spain

Sadaf Jamal Gilani
Department of Pharmaceutical Chemistry
Nibha Institute of Pharmaceutical Sciences
Bihar, India

Syed Sarim Imam
Department of Pharmaceutics
College of Pharmacy
King Saud University
Riyadh, Saudi Arabia

Imogen Ingram
International POPs Elimination
  Network (IPEN)
Rarotonga, Cook Islands

Isangedighi Asuquo Isangedighi
Department of Fisheries and Aquatic
  Environmental Management
University of Uyo
Uyo, Akwa Ibom State, Nigeria

Mohammed Asadullah Jahangir
Department of Pharmaceutics
Nibha Institute of Pharmaceutical Sciences
Bihar, India

Prathmesh Kenjale
Department of Pharmaceutics
Poona College of Pharmacy
Bharati Vidyapeeth University
Maharashtra, India

A. L. Lusher
Department of Animal Ecology I
University of Bayreuth
Bayreuth, Germany

Piyush Mehta
Department of Quality Assurance
Poona College of Pharmacy
Bharati Vidyapeeth University
Maharashtra, India

Abdul Muheem
Dolcera Corporation
Telangana, India

A. Muhtadi
Department of Aquatic Resources
    Management
Faculty of Agriculture
Universitas Sumatera Utara
Medan, North Sumatera, Indonesia

Angel Nadal
Endocrine Society EDC Advisory Group
    Chair
Miguel Hernandez University of Elx
Alacant, Spain

M. T. L. Nascimento
Departamento de Biologia da Faculdade
    de Ciências
and
Centro Interdisciplinar de Investigação
    Marinha e Ambiental (CIIMAR/CIMAR)
Universidade do Porto
Porto, Portugal
and
Departamento de Geologia
Instituto de Geociências da Universidade
    Federal Fluminense
and
Departamento de Engenharia Sanitária
    e do Meio Ambiente da Faculdade de
    Engenharia
Universidade do Estado do Rio de Janeiro
Rio de Janeiro, Brasil

J. A. Baptista Neto
Departamento de Geologia
Instituto de Geociências da Universidade
    Federal Fluminense
Rio de Janeiro, Brasil

Leo M. L. Nollet (retired)
University College Ghent
Ghent, Belgium

Ofonmbuk Ime Obot
Department of Fisheries and Aquatic
    Environmental Management
University of Uyo
Uyo, Akwa Ibom State, Nigeria

M. de Oliveira e Sá
Departamento de Biologia da Faculdade
    de Ciências
and
Centro Interdisciplinar de Investigação
    Marinha e Ambiental (CIIMAR/CIMAR)
Universidade do Porto
Porto, Portugal

R. Pereira
Departamento de Biologia da Faculdade
    de Ciências
and
Centro Interdisciplinar de Investigação
    Marinha e Ambiental (CIIMAR/CIMAR)
Universidade do Porto
Porto, Portugal

Mohammad Rashid
Department of Saidla (Unani
    Pharmaceutics), AKTC
Aligarh Muslim University
Aligarh, India

Asif Raza
Department of Chemistry
Aligarh Muslim University
Aligarh, India

Dolores Romano
Independent Consultant
Zaragoza, Spain

David Santillo
Greenpeace Research Laboratories
Exeter, United Kingdom

A. D. O. Santos
Departamento de Biologia da Faculdade
    de Ciências
and
Centro Interdisciplinar de Investigação
    Marinha e Ambiental (CIIMAR/CIMAR)
Universidade do Porto
Porto, Portugal

and

Departamento de Geologia
Instituto de Geociências da Universidade
    Federal Fluminense
and
Departamento de Engenharia Sanitária
    e do Meio Ambiente da Faculdade de
    Engenharia
Universidade do Estado do Rio de Janeiro
Rio de Janeiro, Brasil

Shariq Shamsi
Department of Ilmul Saidla (Unani
    Pharmaceutics)
National Institute of Unani Medicine
Bangalore, India

Faisal Zia Siddiqui
Department of Chemical Engineering
Faculty of Engineering and
    Technology
Aligarh Muslim University
Aligarh, India

Khwaja Salahuddin Siddiqi
Department of Chemistry
Aligarh Muslim University
Aligarh, India

P. Sobral
MARE – Marine and Environmental
    Sciences Centre
Faculdade de Ciências e Tecnologia
Universidade NOVA de Lisboa
Caprarica, Portugal

Joao Sousa
Global Marine and Polar Programme
International Union for Conservation of
    Nature (IUCN)
Gland, Switzerland

M. N. Vieira
Departamento de Biologia da Faculdade
    de Ciências
and
Centro Interdisciplinar de Investigação
    Marinha e Ambiental (CIIMAR/CIMAR)
Universidade do Porto
Porto, Portugal

H. Wahyuningsih
Department of Biology
Faculty of Mathematics and Natural
    Sciences
Universitas Sumatera Utara
Medan, North Sumatera, Indonesia

Roland Weber
POPs Environmental Consulting
Schwäbisch Gmünd, Germany

N. A. Welden
Faculty of Science, Technology,
    Engineering and Mathematics
Open University
Milton Keynes, United Kingdom

Microplastics and Nanoplastics

CHAPTER 1

Microplastics and Nanoplastics
Origin, Fate and Effects

Vividha Dhapte-Pawar, Piyush Mehta and Prathmesh Kenjale

CONTENTS

1.1 INTRODUCTION

Plastic is a synthetic or semi-synthetic material with a varied range of consumer and industrial applications [1]. These synthetic materials are mostly fabricated from polymers built around chains of carbon atoms, mainly with hydrogen, oxygen, sulfur and nitrogen between the spaces [2]. The term "plastic" is derived from the Greek word "plastikos," meaning suitable for molding [1]. Global production of plastics has increased over the past years and is still enduring (Figure 1.1). Because of poor waste management practices, plastic in the environment, as well as oceans, is also increasing rapidly [3]. It is projected that around 60% of all plastics ever made have accumulated in landfills or the natural environment [4].

Nowadays, plastic items are being manufactured in all shapes and sizes, with the micron and submicron ranges considered to be "micro- or nanoplastics," also termed as "primary micro- or nanoplastics." While, on the other hand, in the environment, the breakdown of large plastic items into small micron or nano-sized fragments are termed as "secondary micro- or nanoplastics." They appear in various micro, as well as in nano, forms, such as fragments,

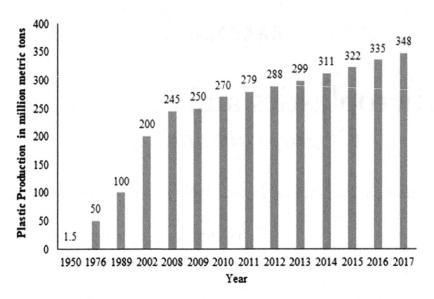

FIGURE 1.1   Global plastic production over the past years (in metric tons).

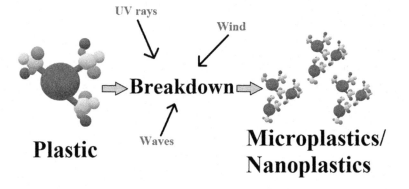

FIGURE 1.2   Formation of micro/nanoplastics by breakdown mechanisms.

fibers or films. Various mechanisms by which this breakdown can occur are mechanical, chemical or biological degradation (Figure 1.2). Mechanical degradation involves tire abrasion, road wear, washing of synthetic textiles, physical weathering of large items [5–8]. Chemical degradation can occur by exposure to acids or alkalis while UV degradation occurs by exposure to non-ionizing, UV radiation. Biological degradation can also occur by the living organisms having capacity to ingest and degrade plastics like waxworms, mealworms and similar microorganisms [9–11]. Additionally, plasticizers added during plastic manufacturing improve durability and flexibility, along with enhancing degradation [12,13].

1.2 MICRO/NANOPLASTIC SOURCES WITHIN THE ENVIRONMENT

Micro/nanoplastics can be released into the environment, either as primary or secondary plastics as described earlier. These plastics can be found on land or in freshwater or marine environments.

1.2.1 Micro/Nanoplastics on Land

The majority of plastics manufactured are being used in consumer products. Increased plastics on land are majorly due to improper waste management, which includes loss during the waste disposal chain, industrial spillages or release from landfill sites. These items can degrade to form secondary micro/nanoplastics within the environment. Micro/nanoplastics may also be released directly to land along with sewage sludge applied to agricultural land as a fertilizer [14–16].

1.2.2 Micro/Nanoplastics in the Freshwater Environment

Freshwaters such as lakes, rivers, ditches, streams and ponds represent the most complex systems regarding micro/nanoplastic transport and retention. They receive plastics from the land and produce micro/nanoplastics through break up of larger items. Larger plastic items can enter the freshwater environment through inadequate waste disposal, either through littering or loss from landfill or through transport from land via wind or surface runoff [17,18].

1.2.3 Micro/Nanoplastics in the Marine Environment

Sources of micro/nanoplastics to marine environments are widespread, as oceans are generally considered to be the ultimate sink for all plastic within the environment. In addition to the inputs from rivers, plastics will also enter oceans directly via mismanaged waste, including abandoned fishing gear, accidental cargo loss and illegal dumping. This will most likely be in the form of plastic waste that will degrade to form micro/nanoplastics within the marine environment. Micro/nanoplastics are prevalent throughout various locations and even within marine organisms worldwide, with ocean currents leading to specific areas of accumulation such as the well-known "Great Pacific Garbage Patch" [19–21].

1.2.4 Microplastics in the Atmosphere

Owing to the lightweight nature of plastic, many micro/nanoplastic particles remain suspended and transported within the air as "urban dust." These commonly originate from road dust (e.g., tire and paint particles) and fibers from synthetic textiles, especially from soft furnishings, and can lead to deposition of microplastics to land or aquatic environments. Although urban dust will originate especially in cities and densely populated areas, air currents and wind can carry plastic particles to other regions far from the source [22–25].

1.3 MICRO/NANOPLASTICS EFFECTS ON PLANTS

Recently, micro/nanoplastics effects in terrestrial ecosystems have come into focus; a decade ago the research was limited to aquatic systems only. Micro/nanoplastics can affect soil properties with consequent influence on plant growth and activity, either positively or negatively. These effects will vary as a function of plant species, plastic type and, thus, are likely to translate to major changes in plant community. Micro/nanoplastics can be viewed conceptually as a soil physical contaminant that may lower soil bulk density and

FIGURE 1.3   Effect of micro/nanoplastics on plants.

enhance soil drying [26–29]. Slow decomposition of plastic particles due to very high carbon content is known to create microbial immobilization [30]. Major effects of the micro/nanoplastics on plants are summarized in Figure 1.3.

1.4 MICRO/NANOPLASTICS EFFECTS ON THE FOOD CHAIN

There is growing concern about the impact of human activities on the whole ecosystem, along with an apprehension that the smaller plastic fraction, through bioaccumulation and trophic transfer, may ultimately contaminate the human population. However, in contrast to the visible presence of massive amounts of litter, there is presently very little understanding about the extent of environmental contamination caused by micro/ nanoplastics. Toxicological effects due to micro/nanoplastics ingestion by humans are still not evident. Presently, it is impossible to assess and control the human exposure to micro- and nanoplastics through food consumption because of the lack of validated methods and standardization on the part of analytical procedures. It must be underlined that the definition of what is intended to be measured is essential for the development of a measurement method, and there is absence of such an internationally accepted definition related to micro/nanoplastic particles. In view of the diversity of particles with respect to size, shape and composition; adsorption of other pollutants and dynamic change of their distribution in our environment based on human activities, the development of fit-for-purpose standardized methods constitutes a challenge [31].

The calibration of such standard methods requires the use of reference materials, where the type of particle and their concentration is precisely known. Different reference materials for diverse polymers in special matrices should be developed. The adsorption of chemicals and microorganisms or the leaching properties of such materials must accurately mimic the adsorption and leaching properties of plastic particles in the natural environment. Then, comparisons between prepared reference materials and naturally occurring particles can be performed (similar to computability studies which compare incurred serum samples to patient samples in a clinical analysis). Finally, to monitor the contamination caused by micro- and nanoplastics, a significant number of samples must be collected, representative of an environmental area or a population, using standard sampling protocols that avoid further contamination. However, no accurate answer can thus be given, neither to the question of how much microplastic we may consume through a normal diet nor to the question of how it may affect us. Until standardized analytical methods and reference materials are available, the data produced will not be representative and significant [31].

1.5 SEPARATION AND ANALYSIS OF MICRO/ NANOPLASTICS IN ENVIRONMENTAL SAMPLES

The vast amount of plastic waste emitted into the environment and the increasing concern of potential harm to wildlife has made micro/nanoplastic pollution a growing environmental concern. Small particles are especially concerning because of their high specific surface area for sorption of contaminants, as well as their potential to translocate in the bodies of organisms. These same small particles are challenging to separate and identify in environmental samples because their size makes handling and observation difficult.

Currently employed passive density and size separation techniques to isolate plastics from environmental samples are not well suited to separate micro/nanoplastics. Passive flotation is hindered by the low buoyancy of small particles as well as the difficulty in handling small particles on the surface of flotation media [32]. These techniques can be improved by adapting active density separation from cell biology and taking advantage of surface-interaction-based separations from analytical chemistry. Furthermore, plastic pollution is often challenging to quantify in complex matrices such as biological tissues and wastewater. Biological and wastewater samples are important matrices that represent key points in the sources and fate, respectively, of plastic pollution. In such samples, protocols need to be optimized to increase throughput, reduce contamination potential and avoid plastics destruction during sample processing.

Post-isolation, micro/nanoplastics from environmental samples should be characterized for both detection and quantification levels. With existing techniques, micro/nanoplastics are difficult to characterize or even detect as their small size and low mass offer restricted signals for visual and spectroscopic estimation. Each of these techniques involve trade-offs in throughput, spatial resolution and sensitivity. Multiple analytical techniques applied in tandem are likely to be required for accurate identification and complete quantification of micro/nanoplastics in environmental samples [32].

1.6 ANALYTICAL METHODOLOGIES TO ASSESS THE MICRO/ NANOPLASTICS WITHIN THE ENVIRONMENT

Analysis of micro/nanoplastics in ecological matrix still remains a difficult task. Range of analytical techniques, tools and procedures are explored for isolation, identification and quantification of plastics. On the basis of diversity in plastics, analytical techniques can vary from physical (visual classification, microscopy) to chemical (spectroscopy, chromatography) techniques. Moreover, few researchers suggest isolation and identification of plastic on the basis of plastic size, density and hydrophobicity. The isolation and identification techniques have significant impact on quantification of plastics from environmental matrix. Typically, in micro/nanoplastics, quantification is highly dependent on the precision and accuracy of the separation techniques [32–34]. Variously commonly used analytical techniques for isolation, identification and quantification of plastics are thoroughly described in Table 1.1, with the key advantages and limitations.

TABLE 1.1 Analytical Techniques for Analyzing and Understanding Micro/Nanoplastics from Different Matrices

Technique	Advantages	Limitations	References
Thermal analysis with mass spectrometry	1. Easy identification of microplastics 2. Identification using thermal degradation behavior 3. No requirement of pretreatment for analysis and require only 5–200 µg of samples 4. Thermal extraction and desorption (TED) with mass spectrometry (MS) permits the simultaneous identification of multiple samples	1. Destructive techniques 2. Do not allow the physical characterization of samples 3. Expensive in nature	[35–37]
Raman spectroscopy	1. Non-destructive spectroscopic method 2. Involve use of rotational and vibrational interactions 3. Provides a structural fingerprint for micro/nanoplastics in different environmental matrices 4. Enables the identification within minutes 5. In combination with other sophisticated analytical techniques (e.g., confocal microscopy, atomic force microscopy), can provide better visual analysis	1. May require pretreatment or purification before sample analysis 2. May need standardization prior to sample analysis 3. Needs specialized staff for analysis and interpretation	[34,38–40]
Infrared spectroscopy	1. Precisely identify the physicochemical interactions between micro/nanoplastics and different matrices 2. Highly specific and reveals characteristic band patterns for individual plastic particles 3. Complementary technique to Raman spectroscopy and vice versa 4. Can easily accommodate other functionalities such as focal plane array (FPA)-based FTIR and attenuated total reflectance (FTIR-ATR) 5. Available as portable machines	1. Time consuming 2. Organic matrices may add to misidentification of plastics spectra 3. Better analysis and outcomes need dry samples	[33,41]
Microscope-assisted visual screening	1. Inexpensive and easy to perform 2. Used for pre-sorting of samples	1. May lead to over- or underestimation 2. No data on chemical arrangement	[42–44]

Plastics' physical (size, shape and morphology) properties and chemical (composition, functional moieties and chemical class) characteristics are the crucial features during analysis. Any analytical technique that consistently measures both the properties is highly appropriate for plastic analysis. However, it is very tricky to achieve identification of both characteristics using a single analytical method. Most of the techniques from Table 1.1 involve analytical techniques which are regularly utilized in physical, mechanical and biological fields. Exploring analytical principles from these domains could assist in the design and development of novel systems to respond to difficulties in micro/nanoplastics analysis. Thus, use of hyphenated analytical techniques (coupling of two or more analytical techniques) is highly applicable. Hyphenated analytical techniques that include elemental analysis and imaging to attain the chemical and physical identification may offer a possible future direction for plastics identification. Apart from these techniques, for most of the chromatographic/spectroscopic methods, we need to develop new and suitable libraries for plastic analysis based on chemical and physical characteristics for optimum identification. These types of libraries will provide quick decoding of the signals during analysis and ease the overall interpretation and data collection process. Simply, it will help to reduce identification time and efforts [32–34].

1.7 THREAT OF MICRO/NANOPLASTICS TO HUMAN BEINGS

It is well documented that plastic particles present in the ecological systems have the potential to affect human beings, directly or indirectly. The judgment of the micro/nanoplastics hazard to human health requires an in-depth understanding of ecological plastics levels and pathways of exposure (Figure 1.4). Plastic particles present in the environment may reach individuals through inhalation of airborne particles, the digestive tract via food products (fish, drink products, beverages and canned food) and via dermal exposure. Micro/nanoplastic particles can also be consumed indirectly via personal care products such as toothpaste and scrubs or even cosmetics. Individuals may also be exposed to plastics through non-ecological resources such as prosthetics and biomedical treatments.

1.7.1 Oral Route

Approximately 50,000 to 15 million plastic particles are discharged into drinking water via various sources, and thus, it is the most common route of plastic exposure to humans [45]. Food products and their packaging materials are another important source of plastic exposure to humans. Aquatic organisms are the major victims of micro/nanoplastic contaminants, either through water or the nourishing material; these organisms may serve as a cause of human plastic exposure. Translocation of microscopic plastic fragments across the gastrointestinal tract of crabs, fish and mussels is well reported [46,47]. Regarding bivalves, individuals eat the complete soft tissues which may include microscopic plastic fragments. Additionally, aquatic creatures may be polluted after collection during their storage and shipping. Human beings may also be directly exposed to micro/nanoplastic particulates from stored (typically weeks to months) food stuffs, which mainly includes harvested and processed animal and plant food items such as packed milk, beverages, vegetables and fruits. The physical existence of plastic particulates is toxic owing to their intrinsic ability to stimulate intestinal blockage or tissue abrasion. Additionally,

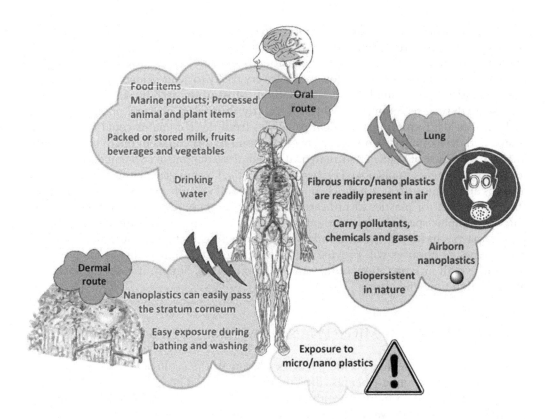

FIGURE 1.4   Pathways of micro/nanoplastics exposure causing hazards to human health.

tissue accumulation, kinetics and the dynamics pattern are mainly governed by micro/nanoplastic particle size and morphology. Apart from the oral route, plastic exposure via dermal and inhalation routes is summarized in the sections below [48].

1.7.2 Dermal Route

Dermal contact may take place when an individual interacts with water polluted with micro/nanoplastic particulates during bathing, washing or via facial/body scrubs containing micro/nanoplastic particulates. However, owing to size of micro/nanoplastic particles, their uptake across skin needs penetration through stratum corneum, which is limited to particles below 100 nm, so absorption via the skin is unlikely to arise. Yet, nanoplastic could easily penetrate into the human skin. Additionally, the excess use of cosmeceuticals or personal care products may also causes harm to human skin [48,49].

1.7.3 Inhalation

Lungs are constantly exposed to surroundings and thus prone to injury caused by pollutants and oxidative stress [48,50]. Human exposure to plastic particulates via the inhalation route could happen after micro/nanoplastics become airborne, typically from wave action in marine environments or during wastewater treatment. Additionally, fragmented fibers microplastics are readily present in polluted air. These fibrous microplastics have the ability to carry harmful pollutants, carcinogens, chemicals and gases. Most of these inhaled fibrous microplastics act as bioperspirants, and their deposition and clearance vary according to their respective concentration, dimensions and morphology. Most of them are subjected to mucociliary clearance; however, a few may persist in the pulmonary airways causing inflammation and other localized biological responses. Overall biological response may worsen, especially for individuals suffering from lung disease. Additionally, these contaminants lead to health effects such as genotoxicity, carcinogenicity and mutagenicity [48,51].

Harmful biological effects from micro/nanoplastic particles may result from a combination of the plastics' inherent properties such as chemical composition, synthetic additives, leaching of components, ability to initiate chemical reactions and, mainly, the non-biodegradable nature. So far, various *in vitro* studies are carried out to highlight toxicity of micro/nanoplastic particles toward human health. Yet, only a few studies are reported using *in vivo* dynamics on plastic [52,53]. Owing to its non-biodegradable nature and chemical composition, it is known to initiate several cellular damages such as alterations of the lysosomal compartment, downregulation of immunological responses, initiation of peroxisomal proliferation and disruption of the antioxidant system. All these consequences may cause an altered gene expression profile and subsequently instigate genotoxicity, mutagenicity and neurotoxicity. Recently, few researchers reported that micro/nanoplastic particles could also act as a vector for pathogens, causing potential distribution in species and disturbing the ecology [54]. It also causes bioaccumulation and dispersion of contaminants across the ecosystem, ultimately threatening human health [48]. Overall, the estimation and isolation of micro/nanoplastic elements from the ecological matrices is a very essential task. Therefore, upcoming investigations should assess and understand the fate of plastic elements with their allied chemicals through the food chain.

1.8 CONCLUSION

The overburden of microplastics and nanoplastics, originating from various sources, is contaminating the environment as well as threatening the health of human beings. Existing techniques of separation and isolation have limitations as submicron plastic particles have high specific surface area for sorption of contaminants along with their potential to translocate in the physiology of organisms via food chain. Hence, more sensitive analytical tools to detect and quantify the presence, identity and amount of microplastics and nanoplastics in ecological samples and food need to be developed, standardized and validated. Standardized quality assurance protocols with uniform legislation to control the dietary exposure should be developed. Parallel concerns on use of nanoplastics in the biomedical and pharmaceutical fields also need to be addressed in the future, with more research on the toxicokinetics and safety profiling.

REFERENCES

1. What are Plastics? https://www.plasticseurope.org/en/about-plastics/what-are-plastics
2. Plastics, https://www.explainthatstuff.com/plastics.html
3. Global plastic production from 1950 to 2017 (in million metric tons), https://www.statista.com/statistics/282732/global-production-of-plastics-since-1950/
4. Geyer R, Jambeck JR, and Law KL. Production, use and fate of all plastics ever made. *Sci Adv* 2017, 3(7), e1700782.

5. Horton AA, Svendsen C, Williams RJ, Spurgeon DJ, and Lahive E. Large microplastic particles in sediments of tributaries of the River Thames, UK - Abundance, sources and methods for effective quantification. *Mar Pollut Bull* 2017, 114(1), 218–26.

6. Hernandez E, Nowack B, and Mitrano DM. Synthetic textiles as a source of microplastics from households: A mechanistic study to understand microfiber release during washing. *Environ Sci Technol* 2017, 51(12), 7036–46.

7. Napper IE and Thompson RC. Release of synthetic microplastic plastic fibres from domestic washing machines: Effects of fabric type and washing conditions. *Mar Pollut Bull* 2016, 112(1), 39–45.

8. Rillig MC. Microplastic in terrestrial ecosystems and the soil? *Environ Sci Technol* 2012, 46(12), 6453–4.

9. Yang J, Yang Y, Wu WM, Zhao J, and Jiang L. Evidence of polyethylene biodegradation by bacterial strains from the guts of plastic-eating waxworms. *Environ Sci Technol* 2014, 48(23), 13776–84.

10. Yang Y, Yang J, Wu WM, Zhao J, Song Y, Gao L, Yang R, and Jiang L. Biodegradation and mineralization of polystyrene by plastic-eating mealworms: Part 1. Chemical and physical characterization and isotopic tests. *Environ Sci Technol* 2015, 49(20), 12080–6.

11. Gu J-D. Microbiological deterioration and degradation of synthetic polymeric materials: Recent research advances. *Int Biodeterior Biodegrad* 2003, 52(2), 69–91.

12. Cole M, Lindeque P, Halsband C, and Galloway TS. Microplastics as contaminants in the marine environment: A review. *Mar Pollut Bull* 2011, 62(12), 2588–97.

13. Talsness CE, Andrade AJ, Kuriyama SN, Taylor JA, and vom Saal FS. Components of plastic: Experimental studies in animals and relevance for human health. *Philos Trans R Soc Lond B Biol Sci* 2009, 364(1526), 2079–96.

14. Lechner A and Ramler D. The discharge of certain amounts of industrial microplastic from a production plant into the river Danube is permitted by the Austrian legislation. *Environ Pollut* 2015, 200, 159–60.

15. Sadri SS and Thompson RC. On the quantity and composition of floating plastic debris entering and leaving the Tamar estuary, Southwest England. *Mar Pollut Bull* 2014, 81(1), 55–60.

16. Nizzetto L, Futter M, and Langaas S. Are agricultural soils dumps for microplastics of urban origin? *Environ Sci Technol* 2016, 50(20), 10777–9.

17. Jambeck J, Geyer R, Wilcox C, Siegler TR, Perryman M, Andrady AL, Narayan R, and Law KL. Plastic waste inputs from land into the ocean. *Science* 2015, 347(6223), 768–71.

18. Imhof HK, Ivleva NP, Schmid J, Niessner R, and Laforsch C. Contamination of beach sediments of a subalpine lake with microplastic particles. *Curr Biol* 2013, 23(19), 867–8.

19. Law KL and Thompson RC. Microplastics in the seas. *Science* 2014, 345(6193), 144–5.

20. Zhang Y, Zhang YB, Feng Y, and Yang XJ. Reduce the plastic debris: A model research on the great Pacific Ocean garbage patch. *Adv Mat Res* 2010, 113, 59–63.

21. Browne MA, Crump P, Niven SJ, Teuten E, Tonkin A, Galloway T, and Thompson R. Accumulation of microplastic on shorelines woldwide: Sources and sinks. *Environ Sci Technol* 2011, 45(21), 9175–9.

22. Dris R, Gasperi J, Saad M, Mirande C, and Tassin B. Synthetic fibers in atmospheric fallout: A source of microplastics in the environment? *Mar Pollut Bull* 2016, 104(1–2), 290–3.

23. Dehghani S, Moore F, and Akhbarizadeh R. Microplastic pollution in deposited urban dust, Tehran metropolis, Iran. *Environ Sci Pollut Res* 2017, 24(25), 20360–71.
24. Dris R, Gasperi J, Mirande C, Mandin C, Guerrouache M, Langlois V, and Tassin B. A first overview of textile fibers, including microplastics, in indoor and outdoor environments. *Environ Pollut* 2017, 221, 453–8.
25. Zylstra ER. Accumulation of wind-dispersed trash in desert environments. *J Arid Environ* 2013, 89, 13–5.
26. de Souza Machado AA, Kloas W, Zarfl C, Hempel S, and Rillig MC. Microplastics as an emerging threat to terrestrial ecosystems. *Glob Chang Biol* 2018, 24(4), 1405–16.
27. de Souza Machado AA, Lau CW, Till J, Kloas W, Lehmann A, and Becker Rillig MC. Impacts of microplastics on the soil biophysical environment. *Environ Sci Technol* 2018, 52(17), 9656–65.
28. Zimmermann RP and Kardos LT. Effect of bulk density on root growth. *Soil Sci* 1961, 91(4), 280–8.
29. Wan Y, Wu C, Xue Q, and Hui X. Effects of plastic contamination on water evaporation and desiccation cracking in soil. *Science of The Total Environment*, 2019, 654, 576–82.
30. Rillig MC. Microplastic disguising as soil carbon storage. *Environ Sci Technol* 2018, 52(11), 6079–80.
31. Toussaint B, Raffael B, Angers-Loustau A, Gilliland D, Kestens V, Petrillo M, Rio-Echevarria IM, and Van den Eede G. Review of micro- and nanoplastic contamination in the food chain. *Food Addit Contam Part A Chem Anal Control Expo Risk Assess* 2019, 36(5), 639–73.
32. Nguyen B, Claveau-Mallet D, Hernandez LM, Xu EG, Farner JM, and Tufenkji N. Separation and analysis of microplastics and nanoplastics in complex environmental samples. *Acc Chem Res* 2019, 52(4), 858–66.
33. Costa J, Reis V, Paço A, Costa M, Duarte AC, and Rocha-Santos T. Micro(nano) plastics – Analytical challenges towards risk evaluation. *Trends Anal Chem* 2019, 111, 173–84.
34. Shim WJ, Hong SH, and Eo S. Identification methods in microplastic analysis: A review. *Anal Methods* 2017, 9, 1384–91.
35. David J, Steinmetz Z, Kučerík J, and Schaumann GE. Quantitative analysis of poly(ethylene terephthalate) microplastics in soil via thermo gravimetry mass spectrometry. *Anal Chem* 2018, 90(15), 8793–9.
36. Silva AB, Bastos AS, Justino CIL, da Costa JP, Duarte AC, and Rocha-Santos TAP. Microplastics in the environment: Challenges in analytical chemistry - a review. *Anal Chim Acta* 2018, 1017, 1–19.
37. Dümichen E, Eisentraut P, Bannick CG, Barthel AK, Senz R, and Braun U. Fast identification of microplastics in complex environmental samples by a thermal degradation method. *Chemosphere* 2017, 174, 572–84.
38. Imhof HK, Schmid J, Niessner R, Ivleva NP, and Laforsch C. A novel, highly efficient method for the separation and quantification of plastic particles in sediments of aquatic environments. *Limnol Oceanogr Methods* 2012, 10, 524–37.
39. Cole M, Lindeque P, Fileman E, Halsband C, Goodhead R, Moger J, and Galloway TS. Microplastic ingestion by zooplankton. *Environ Sci Technol* 2013, 47(12), 6646–55.
40. Käppler A, Fischer D, Oberbeckmann S, Schernewski G, Labrenz M, Eichhorn KJ, and Voit B. Analysis of environmental microplastics by vibrational microspectroscopy: FTIR, Raman or both? *Anal Bioanal Chem* 2016, 408(29), 8377–91.

41. Löder MGJ and Gerdts G. Methodology used for the detection and identification of microplastics—a critical appraisal. In: Bergmann M, Gutow L, and Klages M (eds) *Marine Anthropogenic Litter*. Springer, Cham, 2015, pp. 201–27.

42. Renner G, Schmidt TC, and Schram J. Analytical methodologies for monitoring micro(nano)plastics: Which are fit for purpose? *Curr Opin Environ Sci Health* 2018, 1, 55–61.

43. Wesch C, Barthel AK, Braun U, Klein R, and Paulus M. No microplastics in benthic eelpout (Zoarces viviparus): An urgent need for spectroscopic analyses in microplastic detection. *Environ Res* 2016, 148, 36–8.

44. Galgani F, Hanke G, Werner S, Oosterbaan L, Nilsson P, Fleet D, Kinsey S, Thompson R, van Franeker J, and Vlachogianni T. Guidance on Monitoring of Marine Litter in European Seas, European Commission. *Joint Research Centre* 2013. MSFD Technical Subgroup on Marine Litter (TSG-ML), JRC Technical Report. EUR83985. URL: http://mcc.jrc.ec.europa.eu/documents/201702074014.pdf.

45. Mason SA, Garneau D, Sutton R, Chu Y, Ehmann K, Barnes J, Fink P, Papazissimos D, and Rogers DL. Microplastic pollution is widely detected in US municipal wastewater treatment plant effluent. *Environ Poll* 2016, 218, 1045–54.

46. Browne MA, Dissanayake A, Galloway TS, Lowe DM, and Thompson RC. Ingested microscopic plastic translocates to the circulatory system of the mussel, *Mytilus edulis* (L). *Environ Sci Technol* 2008, 42, 5026–31.

47. Watts JR, Urbina MA, Goodhead R, Moger J, Lewis C, Galloway TS. Effect of microplastic on the gills of the shore crab *Carcinus maenas*. *Environ Sci Technol* 2016, 50, 5364–9.

48. Revel M, Châtel A, and Mouneyrac C. Micro(nano)plastics: A threat to human health? *Curr Opin Environ Sci Health* 2018, 1, 17–23.

49. Sykes EA, Dai Q, Tsoi KM, Hwang DM, and Chan WC. Nanoparticle exposure in animals can be visualized in the skin and analyzed via skin biopsy. *Nat Commun* 2014, 5, 3796.

50. Mehta P, Bothiraja C, Mahadik K, Kadam S, and Pawar A. Phytoconstituent based dry powder inhalers as biomedicine for the management of pulmonary diseases. *Biomed Pharmacother* 2018, 108, 828–37.

51. Gasperi J, Wright SL, Dris R, Collard F, Mandin C et al. Microplastics in air: Are we breathing it in? *Current Opinion in Environmental Science & Health* 2018, 1, 1–5.

52. Schirinzi GF, Pérez-Pomeda P, Sanchís J, Rossini C, Farré M, and Barceló D. Cytotoxic effects of commonly used nanomaterials and microplastics on cerebral and epithelial human cells. *Environ Res* 2017, 159, 579–87.

53. Powell JJ, Thoree V, and Pele LC. Dietary microparticles and their impact on tolerance and immune responsiveness of the gastrointestinal tract. *Br J Nutr* 2007, 98, S59–63.

54. Zettler ER, Mincer TJ, and Amaral-Zettler LA. Life in the "Plastisphere"; microbial communities on plastic marine debris. *Environ Sci Technol* 2013, 47(13), 7137–46.

CHAPTER 2

Plastic Waste in the Aquatic Environment
*Impacts and Management**

*Isangedighi Asuquo Isangedighi, Gift Samuel David
and Ofonmbuk Ime Obot*

CONTENTS

2.1 INTRODUCTION

Plastics are synthetic polymers, and they have only existed for just over a century. Plastic is a material consisting of any of a wide range of synthetic or semi-synthetic organic compounds that are malleable and can be molded into solid objects [1]. Plastics are produced by the conversion of natural products or by synthesis from primary chemicals, generally from oil, natural gas or coal [2]. Plastics are typically organic polymers of high molecular mass, but they often contain other substances. Most plastics are made from synthetic resins (polymers) through the industrial process of polymerization [1].

In contemporary society, plastic has attained a pivotal status, with extensive commercial, industrial, medicinal and municipal applications [3]. It ranks among the most widely used materials in the world [4]. In the last 60 years, plastic has become a useful and versatile material with a wide range of applications [5,6]; see Table 2.1. Over 300 million tons of new plastics are used every year. Half of these are used just once and usually for less than 12 minutes. Eight million tons of plastic waste ends up in the ocean every year. So much is getting into the ocean that in some places, these plastic particles outnumber plankton by a ratio of 26:1 [7]. A UNEP report [8] grossly underestimated the amount of plastic debris entering the environment at 8 million pieces per day. However, it should be noted that none of these estimates are licensed to any particular source and should be treated with caution. The world's oceans are diverse and immense; hence, to get a handle on the estimated average level of plastic waste is a very difficult task. About 49% of all produced plastics are buoyant, which gives them the ability to float, and thereby travel on ocean currents to any place in the world [9]. Figure 2.1 show how plastics are littered on the aquatic environment. Annual plastic production has increased dramatically from 1.5 million tons in the 1950s to approximately 280 million tons in 2011 [5]. Annual plastic production is estimated to hit 400 million tons by the year 2020 [10]; this is depicted in Figure 2.2. The most common plastics are polyethylene (PE), polyethylene terephthalate (PET) and polypropylene (PP) [10].

TABLE 2.1 Types of Plastics and Common Use

Polymer Type	Examples
Polyethylene terephthalate	Fizzy drink and water bottles, salad trays
High density polyethylene	Milk bottles, bleach, cleaners and most shampoo bottles
Polyvinyl chloride	Pipes, fittings, window and door frames (rigid PVC), thermal insulation (PVC foam) and automotive parts
Low density polyethylene	Carrier bags, bin liners and packaging films
Polypropylene	Margarine tubs, microwaveable meal trays, also produced as fibers and filaments for carpets, wall coverings and vehicle upholstery
Polystyrene	Yoghurt pots, foam hamburger boxes and egg cartons, plastic cutlery, protective packaging for electronic goods and toys. Insulating material in the building and construction industry.
Unallocated references	Polycarbonate, which is often used in glazing for the aircraft industry

Source: Adapted and modified from Koushal V et al. *Int J Waste Resour* 2014, 4(1), 134–139.

FIGURE 2.1 Examples plastic litter in the aquatic environment: From left to right, plastic debris on the ocean seabed, surface water and beach. (From Lytle GM. When the mermaids cry: The great plastic tide. Available on: http://www.plastic-pollution.org, accessed on 4 January 2018.)

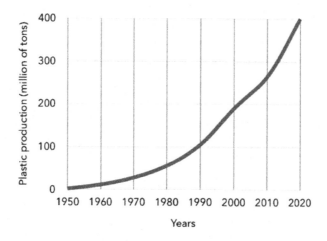

FIGURE 2.2 Evolution of world plastic production, in million tons, from 1950 to 2020. (From Plastics – the Fact 2014. An analysis of European plastics production, demand and waste data. Available online: http://2015.igem.org/wiki/images/e/ef/IGEM_Pasteur_Plastics_the_I, accessed on January 19, 2017.)

According to the Plastic Ocean Foundation [7], the typical characteristics that render plastics so useful relate primarily to the fact that they are both flexible and durable. These characteristics become handy when plastics are used in everyday life; but when they are discarded into the environment, they become a nuisance. Due to their nearly indestructible morphology and the toxins they contain, plastics can seriously affect ecosystems [11]. The biggest mass of plastics debris occurs in the oceans major gyres [12]. Therein, the rotation of vortex centers, where it accumulates currently, the plastics debris patch in the North Pacific Ocean covers an area as large as France and Spain together [12]. This debris affects all ocean life, and because we are at the top of the food chain, it affects humans too [4]. It has now been several decades since the use of plastics exploded, and there is evidence that current approaches to production, use, transport and disposal of plastic materials caused, and are still causing, serious effects on wildlife, and this is not sustainable [4].

The objective of this review is to identify the sources and fate of plastic wastes on the aquatic ecosystem, to examine its impact on the aquatic biota and to suggest ways to ameliorate this problem.

2.2 PLASTIC WASTE IN THE AQUATIC ECOSYSTEM

Plastic now accounts for 10% of all waste generated, with global use exceeding 260 million tons per annum [14]. Plastics waste has accumulated in the environment at an uncontrollable rate where it is subjected to wind and river-driven transport ultimately reaching the coast. Owing to its light weight and durable nature, plastic has become a prevalent, widespread element of marine litter [2,15]. The difficulty in eliminating plastics waste is due to the fact that it does not biodegrade in nature but only photodegrades into smaller pieces.

Because of frequent inappropriate waste management practices, or irresponsible human behavior, large masses of plastic items have been released into the environment and have consequently entered the world's oceans. This process continues and, in some places, is even increasing [4]. Even if plastics are found deep in land, they eventually find their way to the sea or ocean through rivers and streams [16]. The release of plastics into the environment is a result of inappropriate waste management, improper human behavior or incidental pollution [14]. Sources of plastic waste into the aquatic ecosystem include municipal plastic waste (sachet water film bags, etc.), commercial plastic waste (packaging materials, etc.), industrial plastic waste (polyethylene film wrapping, spare parts for cars, PVC pipes and fittings, etc.) and plastic waste from ships [17,18].

Most polymeric materials that enter the environment are subjected to degradation that is caused by a combination of factors, including thermal oxidation, photooxidative degradation, biodegradation and hydrolysis [19]; see Table 2.2. Figure 2.3 shows the pathways and modes of movement for plastic in the marine environment. The common plastics found in the marine environment, however, do not biodegrade and primarily break down through photooxidative degradation. Furthermore, unlike plastics exposed on land, exposed plastics floating on the oceans' surfaces do not suffer from heat buildup due to absorption of infrared radiation, and the reform barely undergo thermal oxidation [20].

The degradation of negatively buoyant plastics depends on very slow thermal oxidation, or hydrolysis, as a result of most wavelengths being readily absorbed by water. Hence, plastics residing in marine environments degrade at a significantly slower rate than they do on land. Polymer degradation can be categorized as any physical or chemical changes resulting from environmental factors, including light, heat, moisture, chemical conditions and biological activity [19]. It is not clear just how long plastic items remain in their original form. However, some plastic items appear to be broken up into smaller and smaller fragments over time [21]. At sea, this process is thought to occur because of wave action, oxidation and ultraviolet light. On the shore, plastic may break up into smaller

TABLE 2.2 Possible Degradation Routes of Synthetic Polymers

Requirements	Photo-Degradation	Thermal/Thermo-Oxidative Degradation	Biodegradation
Active ingredients	UV-light/high-energy radiation	Heat and/or oxygen	Microbial Agents
Heat requirement	None	Above ambient temperature	None
Degradation rate	Slow initiation, fast propagation	Fast	Moderate

Source: From Cooper DA. Effects of chemical and mechanical weathering processes on the degradation of plastic debris on marine beaches. PhD Thesis, The University of Western Ontario, Ontario, 2012.

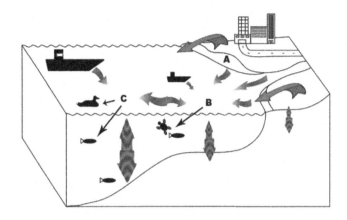

FIGURE 2.3 Schematic diagram indicating the pathways and modes of movement for plastic in the marine environment. (a) Beaches, (b) coastal waters and sediments and (c) open ocean.

pieces owing to grinding from rocks and sand [22]. The resulting plastic fragments may be mistaken for prey and ingested by marine organisms. Plastic debris in the oceans may eventually be broken up so much that it becomes microscopic in size like grains of sand, hence the term microplastics. These tiny fragments (about 20 μm in diameter) have been identified in marine sediments and in ocean waters [23].

Various studies have demonstrated occurrence of marine debris worldwide, with plastic making up the major proportion of marine debris. Reports from some of these studies are summarized in Table 2.3. Though the methods were not assessed to ensure that the results were comparable, Table 2.3 clearly indicates the predominance of plastics among the marine litter, and its proportion consistently varies between 60% and 80% of the total marine debris [24].

Green arrows indicate wind-blown debris, gray arrows display vertical movement through the water column including sea floor deposition and burial, black arrows point to ingestion and entanglement by marine organisms and red arrows indicate waterborne debris [25].

2.3 IMPACTS OF PLASTIC WASTE ON THE BIOTA OF THE AQUATIC ECOSYSTEM

The properties that make plastics such desirable materials for modern society can make them lethal for wildlife when introduced into the environment [4]. Numerous species are affected by plastic pollution, primarily because organisms become entangled in plastic nets, or plastic objects are ingested when organisms mistake plastic debris for food [54]. Another problem of plastics pollution is that it facilitates the transport of species to other regions, alien species hitchhike on floating debris and invade new ecosystems, thereby causing a shift in species composition or even extinction of other species [55]. Plastics also transfer contaminants to the environment or, when ingested, to organisms [56]. A technical report considering the impacts of marine debris on biodiversity revealed that over 80% of reported incidents between organisms and marine debris were associated with plastic while 11% of all reported encounters were with microplastics [57]. Although plastic is often buoyant, it can sink to the bottom of the sea, pulled down by certain "bottom-hugging" currents, oceanic fronts or rapid and heavy fouling. Sediment may

TABLE 2.3   Proportion of Plastics among Marine Debris Worldwide (per no. of items)

Locality	Litter Type	Percentage of Debris Items Represented by Plastics (%)	Source
1992 International Coastal Cleanups	Shoreline	59	[26]
St. Lucia, Caribbean	Beach	51	[27]
Dominica, Caribbean	Beach	36	[27]
Curacao, Caribbean	Beach	40/64	[28]
Bay of Biscay, NE Atlantic	Seabed	92	[29]
NW Mediterranean	Seabed	77	[30]
French Mediterranean Coast	Deep sea floor	>70	[31]
European coasts	Sea floor	>70	[32]
Caribbean coast of Panama	Shoreline	82	[33]
Georgia, US	Beach	57	[34]
5 Mediterranean beaches	Beach	60–80	[35]
50 South African beaches	Beach	>90	[24]
88 sites in Tasmania	Beach	65	[24]
Argentina	Beach	37–72	[24]
9 Sub-Antarctic islands	Beach	51–88	[24]
South Australia	Beach	62	[24]
Kodiak Is., Alaska	Seabed	47–56	[36]
Tokyo Bay, Japan	Seabed	80–85	[37]
North Pacific Ocean	Surface waters	86	[38]
Mexico	Beach	60	[39]
Transkei, South Africa	Beach	83	[40]
National Parks in US	Beach	88	[41]
Mediterranean Sea	Surface waters	60–70	[42]
Cape Cod, US	Beach/harbor	90	[43]
4 North Atlantic harbors, US	Harbor	73–92	[43]
Is. Beach State Park, New Jersey, US	Beach	73	[44]
Halifax Harbor, Canada	Beach	54	[45]
Prince Edward Is., Southern Ocean	Beach	88	[46]
Gough Is., Southern Ocean	Beach	84	[46]
Heard Is., Southern Ocean	Beach	51	[47]
Macquarie Is., Southern Ocean	Beach	71	[47]
New Zealand	Beach	75	[48]
Two gulfs in W. Greece	Seabed	79–83	[49]
South German Bight	Beach	75	[50]
Bird Is., South Georgia, Southern Ocean	Beach	88	[51]
Fog Bay, N. Australia	Beach	32	[52]
South Wales, UK	Beach	63	[53]

Source: From Derraik JGB. *Mar Pollut Bull* 2002, 44(9), 842–852.
Abbreviation: Is., Island.

FIGURE 2.4 Marine organisms ingesting plastic debris: From left to right, recovering plastic debris from carcass of an albatross, plastic debris retrieved from a filleted fish and a turtle eating plastics (polyethylene sachet film). (From Lytle GM. When the mermaids cry: The great plastic tide. Available on: http://www.plastic-pollution.org, accessed on January 4, 2018.)

FIGURE 2.5 Entanglement effect of plastic debris on seals: From left to right, seal trapped by a discarded fishing net and seal entangled by ghost fishing lines. (From Lytle GM. When the mermaids cry: The great plastic tide. Available on: http://www.plastic-pollution.org, accessed on January 4, 2018.)

also help keep plastic on the seafloor [5]. It is likely that once on the seafloor, plastic will change the workings of the ecosystem. Goldberg [58] has suggested the plastic sheets could act like a blanket, inhibiting gas exchange and leading to anoxia or hypoxia (low oxygen levels). Plastic waste could also create artificial sand grounds [59] and cause problems, especially for burying creatures.

There is still relatively little information on the impact of plastics pollution on ocean ecosystems [60,61]. There is, however, an increasing knowledge about their deleterious impacts on marine biota [58]. The threats to marine life are primarily mechanical and are due to ingestion of plastic debris and entanglement in packaging bands synthetic ropes and lines or drift nets [38,54,60]; see Figures 2.4 and 2.5. Since the use of plastics continues to increase, so does the amount of plastics polluting the marine environment. In addition to the impact on marine environment and life, plastic debris can also damage the marine industry (entangling propellers and blocking cooling systems). It has been estimated that marine debris damage to the marine industry in the Asia-Pacific region coasts $1.26 billion annually [62].

2.3.1 Impact on Faunal Communities

The magnitude of plastic pollution carried to sea has significantly multiplied over the past several decades. Oftentimes, wildlife is injured owing to entanglement or ingestion of the plastics found in the environment. It was shown that at least 267 marine species worldwide suffer from entanglement and ingestion of plastic debris [54]; see Table 2.4. When such contact occurs, organisms are seriously affected in a way that often results in death. It is very difficult to estimate what the total effect of plastic debris in the ocean is or to predict the consequences for organisms that ingest or otherwise contact that debris because this cannot be directly observed. By contrast, entanglement can be observed, and it is the most visible effect of plastic debris on organisms in the marine environment. Laist [54] studied and composed a comprehensive list of species that suffered from entanglement caused by marine debris entanglement (see Table 2.4). Nevertheless, the exact extent of entanglement faced by marine organisms is difficult to qualify, because entanglement generally occurs in areas remote from human activity.

Entanglement can cause death by drowning, suffocation, strangulation or starvation [21]. Very often, birds, small whale species and seals drown in ghost nets. They may also lose their ability to catch food or cannot avoid predators because of their entanglement [1].

TABLE 2.4   Number and Percentage of Marine Species Worldwide with Documented Entanglement and Ingestion Records

Species Group	Total Number of spp. Worldwide	Number and % of spp. with Entanglement Records	Number and % of spp. with Ingestion Records
Sea turtles	7	6 (86%)	6 (86%)
Seabirds	312	51 (16%)	111 (36%)
Penguins (Sphenisciformes)	16	6 (38%)	1 (6%)
Grebes (Podicipediformes)	19	2 (10%)	0
Albatrosses, petrels and shearwaters (Procellariiformes)	99	10 (10%)	62 (63%)
Pelicans, boobies gannets, cormorants, frigate birds and tropicbirds (Pelicaniformes)	51	11 (22%)	8 (16%)
Shorebirds, skuas, gulls, terns, auks (Charadriiformes)	122	22 (18%)	40 (33%)
Other birds	–	5	0
Marine mammals	115	32 (28%)	26 (23%)
Baleen whales (Mysticeti)	10	6 (60%)	2 (20%)
Toothed whales (Odontoceti)	65	5 (8%)	21 (32%)
Fur seals and sea lions (Otariidae)	14	11 (79%)	1 (7%)
True seals (Phocidae)	19	8 (42%)	1 (5%)
Manatees and dugongs (Sirenia)	4	1 (25%)	1 (25%)
Sea otter (Mustellidae)	1	1 (100%)	0
Fish	–	34	33
Crustaceans	–	8	0
Squid	–	0	1
Species total		136	137

Source: From Laist DW. Impacts of marine debris: Entanglement of marine life in marine debris including a comprehensive list of species with entanglement and ingestion records. In: Coe JM, and Rogers DB (eds) *Marine Debris: Sources, Impacts, Solutions.* Springer-Verlag, New York, USA, 1997, pp. 99–139.

Plastic debris that pollutes the marine environment is often ingested by marine birds, mammals, turtles and fish [54]. The ingestion of plastics primarily occurs when it is mistaken for food, but can also occur from incidental intake. The ingested materials often consist of micro- and meso-sized debris fragments, which sometimes are able to pass through the gut without hurting the organisms. In most cases, however, fragments become trapped inside the stomach, throat or digestive tract and cause damage (e.g., sharp objects) or a false sense of fullness, which will result in starvation. Table 2.4 shows the list of affected species by plastic waste. The list of affected species indicates that marine debris is affecting a significant number of species [54]; see Table 2.4. It affects at least 267 species worldwide, including 86% of all sea turtle species, 44% of all seabird species and 43% of all marine mammal species [54]. The problem may be undiscovered over vast ocean areas, as the affected animals either sink or are eaten by predators [63]. Globally at least 23% of marine mammal species, 36% of seabird species and 86% of sea turtle species are known to be affected by plastic debris [64].

2.3.1.1 Coastal and Marine Birds

Many birds in the marine environment dive for food and thereby come into contact with plastic debris. The greatest causes of entanglement by seabirds are fishing lines and six-pack rings. Both materials are often transparent and difficult to see. If seen, they can be mistaken for jellyfish and other food [21]. The gannet is one marine bird species that is endangered by plastic debris. As a "plunge-diver," the gannet dives from great heights into the ocean and can thereby be caught by ghost nets or other debris. A study at the Island of Helgoland in Germany, which hosts a large gannet colony, showed that between 1976 and 1985, 29% of dead gannets found had become entangled in net fragments [65]. High proportions of coastal and marine bird species (36% of the 312 species worldwide) ingest plastic fragments [54]; see Figure 2.4. Allsopp et al. [21] reported that 111 out of 312 species of seabirds are known to have ingested debris, and this can affect a large percentage of a population (up to 80%). Moreover, plastic debris is also known to be passed to the chicks in regurgitated food from their parents.

One harmful effect from plastic ingestion in birds is weight loss due, for example, to a falsely sated appetite and failure to put on adequate fat stores for migration and reproduction [21]. Although plastics are mainly ingested by birds because they are mistaken for food, they may also already be present in the gut of their prey or may be passed from adult to chick by regurgitation feeding. Some species feed selectively on plastics fragments that have a specific shape or color [66]. Robards et al. [67] examined the gut content of thousands of birds in two separate studies and found that the ingestion of plastics by seabirds had significantly increased during the 10–15 years between the studies. A study done in the North Pacific by Blight and Burger [68] found plastics particles in the stomachs of 8 of the 11 seabird species caught as by-catch. Harmful effects from the ingestion of plastics include blockage of gastric enzyme secretion, diminished feeding stimulus, lowered steroid hormone levels, delayed ovulation and reproductive failure [69]. The food uptake causes internal injury and death following blockage of the intestinal tract [70–73]; the extent of the harm, however, varies among species. Laist [38] and Fry et al. [74] observed that adults that manage to regurgitate plastic particles could pass them onto the chicks during feedings.

The harm from ingestion of plastic is nevertheless not restricted to seabirds. For Procellariformes such as the albatrosses (Figure 2.2), shearwaters or petrels, the appearance of eroded plastic pieces are similar to many types of food they consume [68]. Small plastics such as bottle caps are often mistaken by seabirds (Procellariformes) for food. In several studies, it was found that diving birds that fed on fish in the water column had less plastic in their stomachs compared with those that were surface eaters [68,75]. This could be because birds that maintain a diet of zooplankton may not be able to distinguish between plastics and their primary source of food owing to the color or shape of the plastic pieces [76]. Since most adult birds regurgitate what has been ingested as a way to feed their chicks, they pass the bolus containing the plastic to the first young. Birds such as the albatross and shearwater had more plastic in the first region of their stomachs and gizzards, indicating that when these plastics were regurgitated, they would be passed to their young during feeding [66]. Juvenile albatrosses and shearwaters were found to ingest more plastics than adults [76,77]. Similar to other marine life, swallowed plastics can obstruct and damage a bird's digestive system, reducing its foraging capabilities. Ryan [72] concluded that ingested plastics could reduce the fitness, growth rate and food consumption of seabirds. Based on the results form a study using domestic chickens (*Gallus domesticus*), the amount of plastic ingested by different species of birds may be an indicator of the accumulation of plastics in an area.

2.3.1.2 Fish

The incidence of accidental entanglement of fish species is difficult to estimate, because certain fish are "intended" to become entangled in nets. Therefore, research emphasizes by-catch of endangered species. For example, between 1978 and 2000, 28,687 sharks were caught in nets that protected people at popular swimming beaches in Kwazulu, South Africa [4]. Microplastics resemble phytoplankters, which are eaten by fish and cetaceans [78]; see Figure 2.4. Ingested plastic debris has been found to reduce stomach capacity, hinder growth, cause internal injuries and create intestinal blockage [79].

Ghost fishing can lead to economic losses for fisheries [21]. For example, an experimental study on ghost fishing of monkfish from lost nets in the Cantabrian Sea in northern Spain, estimated that 18.1 tons of monkfish are captured annually by abandoned nets. This represented 1.46% of the commercial landings of monkfish in the Cantabrian Sea [80]. A study on ghost fishing by lost pots off the coast of Wales, UK, noted that potential losses to the brown crab fishery caused by ghost fishing could be large [81]. In the US, it is estimated that $250 million of marketable lobster is lost annually to ghost fishing [82].

2.3.1.3 Mammals

Many seal species are curious and playful and, especially young seals, are attracted to plastic debris and swim with it or poke their head through loops. Plastic rings, loops or lines easily glide onto the seal's neck, but are difficult to remove owing to the backward direction of the seal's hair (see Figure 2.5). As the seal grows, the plastic collar tightens and strangles the animal [4]. After entanglement in these nets, the animals are not able to reach the water surface, and thus, they drown. An estimated 58% of seal and sea lion species are known to have been affected by entanglement, including the Hawaiian monk seal, Australian sea lions, New Zealand fur seals and species in the Southern Ocean [21].

Whales also become entangled in marine debris. However, although some whale species are incapable of freeing themselves and consequently drown, the larger-sized whales often drag fishing gear away with them. This latter type of entanglement can cause strangulation and can affect the feeding ability of the whale in ways that causes starvation. At least 26 species of cetaceans have been documented to ingest plastic debris [83]. A young male pygmy sperm whale (*Kogia breviceps*) stranded alive in Texas died in a holding tank 11 days later. Stomach compartments were completely occluded by plastic debris, including a garbage can liner, a bread wrapper, a corn chip bag and two other pieces of plastic sheeting [83]. Entanglement is a particular problem for marine mammals, such as fur seals, which are curious when catching food and, to avoid predators, may incur wounds from abrasive or cutting action of attached debris [38,84]. According to Feldkamp et al. [85], entanglement can greatly reduce fitness, as it leads to a significant increase in energetic costs of travel for the northern fur seals (*Callorhinus ursinus*). For instance, they stated that net fragments over 200 g could result in a fourfold increase in the demand of food consumption to maintain body condition [85]. Most cetaceans live far from the shoreline which limits the amount of research on the ingestion of marine debris. If plastic causes unnatural death, cetaceans will most likely sink to the bottom of the ocean [83]. Occasionally, cetaceans will wash ashore, allowing for postmortem examinations. Due to cetacean echolocation capabilities, mistaken consumption of plastic is not probable [86].

Ingestion is most likely because the debris was mixed in with the desired food. Two sperm whales (*Physeter macrocephalus*) were found off the coast of northern California in 2008 with a large amount of fishing gear in their gastrointestinal tracts [87]. One of the sperm whales had a rupture in their compartment of the stomach caused by nylon netting. In the other whale, netting, fishing line and plastic bags were completely blocking the stomach from the intestines [87]. On the coast of Nova Scotia, Canada, a juvenile porpoise (Phocoenidae) was found dead with a balled-up piece of black plastic in the esophagus entangled with three spine stickleback fish [83]. In Brazil, the stomach analysis of Blainville's beaked whale (*Mesoplodon densirostris*) showed the presence of a large bundle of blue plastic thread occupying a substantial part of the stomach chamber [86]. Ayalon et al. [88] argued that within the last decade, at least mass amounts of tangled nylon rope and other debris, including a crayfish pot and a buoy , have been found in the stomachs of whales. Currently, there have not been enough trends found in collected data that prove ingested plastics are the primary cause of death contributing to the decline of cetaceans [89–91]. Plastic entanglement with nets or other materials can result in strangulation, reduction of feeding efficiency and, in some cases, drowning [92]. Because of natural curiosity, pinnipeds often become entangled in marine debris of life [92].

2.3.1.4 Reptiles

Polythene bags drifting in ocean currents look much like the prey items targeted by turtles [93–95]; see Figure 2.4. There is evidence that their survival is being hindered by plastics debris [96], with young sea turtles being particularly vulnerable [97]. Balazs [98] listed 79 cases of turtles whose guts were full of various sorts of plastic debris, and O'Hara et al. [99] cited a turtle found in New York that had swallowed 540 m of fishing line. Entanglement in plastic debris, especially in discarded fishing gear, is a very serious threat to marine reptiles. It also affects the survival of sea turtles as there are endangered [97].

2.3.1.5 Invertebrates

Plastic effects on invertebrates are mainly attributed to microplastics. Various studies have been well documented [3,100–103]. Some of the effects of microplastics on aquatic invertebrates include blockages throughout the digestive system or abrasions from sharp objects resulting in injuries, blockage of enzyme production, diminished feeding stimulus, nutrient dilution, reduced growth rates, lowered steroid hormone levels, delayed ovulation and reproductive failure and absorption of toxins [3]. There is potential for microplastics to clog and block the feeding appendages of marine invertebrates or even to become embedded in tissues [1]. Table 2.5 presents a comprehensive list of evidence of impacts of plastic debris on marine organisms by various studies.

TABLE 2.5 Peer-Reviewed Studies Demonstrating Evidence of Impacts of Plastic Marine Debris

Study	Animal	Encounter Type	Predominant Debris Type	Impact Response
[92]	Grey seals	Entanglement	MF line, net, rope	Constriction
[104]	Manatees	Entanglement	MF line, bags, other debris	Death
[105]	Elephant seals	Entanglement	MF line, fishing jigs	Dermal wound
[106]	Fur seals	Entanglement	Packing bands, fishing gear, other debris	Dermal wound
[107]	Seabirds, pinnipeds	Entanglement	Fishing gear	External wound
[108]	Fur seals	Entanglement	Trawl netting, packing bands	Death
[108]	Fur seals	Entanglement	Trawl netting, packing bands	Reduced population size
[109]	Invertebrates, fish, seabirds, marine mammals	Entanglement	Derelict gillnets	Death
[110]	Seabirds, marine mammals	Entanglement	Plastic, fishing line	Death
[111]	Gorgonians	Entanglement	Fishing line	Damage/breakage
[112]	Sea turtles	Entanglement	Fishing gear	Death
[113]	Whales	Entanglement	Plastic line	Dermal wound
[114]	Whales	Entanglement	Plastic line	Dermal wound
[104]	Manatees	Ingestion	MF line, bags, other debris	Death
[115]	Sea turtles	Ingestion	MF line, fish hooks, other debris	Intestinal blockage, death
[116]	Penguins	Ingestion	Plastic, fishing gear, other debris	Perforated gut, death
[117]	Lugworms (laboratory)	Ingestion	Microplastics	Biochemical/cellular, death
[95]	Sea turtles	Ingestion	Plastic bags, ropes	Gut obstruction, death
[118]	Seabirds	Ingestion	Plastic particles, pellets	Perforated gut
[119]	Fish (laboratory)	Ingestion	Nanoparticles	Biochemical/cellular
[120]	Seabirds	Ingestion	Plastic pellets, foam	Biochemical/cellular

(Continued)

TABLE 2.5 (*Continued*)   Peer-Reviewed Studies Demonstrating Evidence of Impacts of Plastic Marine Debris

Study	Animal	Encounter Type	Predominant Debris Type	Impact Response
[107]	Seabirds, pinnipeds	Ingestion	Fishing hooks	Internal wound
[90]	Sperm whale	Ingestion	Identifiable litter items	Gastric rupture, death
[74]	Seabirds	Ingestion	Plastic fragments, pellets, identifiable litter	Gut impaction, ulcerative lesions
[87]	Sperm whales	Ingestion	Fishing gear, other debris	Gastric rupture, gut impaction, death
[121]	Copepods (laboratory)	Ingestion	Micro- and Nanoplastics	Death
[122]	Fish (laboratory)	Ingestion	Microplastics	Biochemical/cellular
[123–125]	Fish (laboratory)	Ingestion	Microplastics	Biochemical/cellular
[72]	Birds (laboratory)	Ingestion	Microplastics	Reduced organ size
[112]	Sea turtles	Ingestion	Marine debris	Gut obstruction
[3]	Lugworms (laboratory)	Ingestion	Microplastics	Biochemical/cellular
[126]	Mussels (laboratory)	Ingestion and gill uptake	Microplastics	Biochemical/cellular
[127]	Epibenthic megafauna	Interaction (contact)	Plastic bottles, glass jars	Altered assemblage
[128]	Sessile invertebrates (coral reef)	Interaction (contact)	Lobster traps	Altered assemblage
[129]	Assemblage on sediment	Interaction (contact)	Plastic litter	Altered assemblage
[130]	Sessile invertebrates (coral reef)	Interaction (contact)	MF line, lobster trap, hook and line gear	Tissue abrasion
[131]	Sessile invertebrates (coral reef)	Interaction (contact)	Hook and line gear	Tissue abrasion
[132]	Seagrass	Interaction (contact)	Crab pots, tires, wood	Breakage, suffocation, death
[133]	Sea turtles	Interaction (obstruction)	Waste, medical waste	Reduced population size
[134]	Ghost crabs	Interaction (obstruction)	Beach litter, mostly plastic	Reduced population size
[134]	Ghost crabs	Interaction (substrate)	Beach litter, mostly plastic	Altered assemblage
[101]	Marine insects	Interaction (substrate)	Microplastics	Increased population size

Source: From Law KL. *Annu Rev Mar Sci* 2017, 9, 205–229.
Note: This table is based on analysis by [135] for publications through the year 2013, extracting studies for plastic marine debris only.
Abbreviation: MF, monofilament line.

2.3.2 Impact on Floral Communities

Impact on plant communities is minimal compared to threats to animals. Natural flotsam, of both marine and terrestrial origin (seaweeds and plants), together with jetsam of indeterminate sources, tends to accumulate along high tide strandlines, where it is commonly known as the "the wrack" [59]. These areas are often ephemeral, dynamic and seasonal environments and also tend to accumulate significant quantities of manufactured materials, in particular, those also made of plastic and other indestructible materials [59]. As a consequence, wrack environments are commonly unsightly, and the demands of local authorities to clean up the mess are frequent and can be expensive [137,138]. Maximum impacts on the floral communities are, however, observed in the form of microplastics. Recent studies (such as [3]) emphasized the important role of microplastics as they are easily ingestible by small organisms, such as plankton species, and form a pathway for contaminants to enter the food web.

2.3.3 Invasive Species

The presence of industrial pellets on beaches free from the influence of petrochemical facilities and pellet processing plants is an indication of long-range marine transport [139]. Like all natural or artificial floating debris, plastic can provide a mechanism for encrusting and fouling organisms to disperse over great distances [140]. Logs, pumice and other flotsam have traversed the open ocean for millennia [59], and the introduction of hard plastic debris to the marine ecosystem may provide an appealing and alternative substrate for some opportunistic colonizers [59,141]. It is estimated that if biotic mixing occurs, global marine species diversity may decrease by up to 58% [142]. Barnes [143] estimates the propagation of fauna in the sea has doubled in the tropics, and more than tripled at high latitudes (>50°), due to the input of anthropogenic debris. The hard surfaces of plastics provide an ideal substrate for opportunistic colonizers. Pelagic plastics are most commonly colonized by bivalve mollusks; however, other encrusting organisms include bacteria, diatoms, algae and barnacles [1,59,70,134,141].

Plastic substrates may also contain multispecies habitats composed of organisms that would normally inhabit different ecological niches [140]. Drifting plastic debris may also increase the range of certain marine organisms or possibly introduce species to new environments which they had previously not inhabited [144]. Sensitive or at-risk littoral, intertidal and shoreline ecosystems could be negatively affected by the arrival of unwanted and aggressive alien species, with potentially damaging environmental consequences [140,145,146,59]. The absence of biological organisms on plastic debris may be an indication that the particles were not present in the marine environment long enough for fouling to occur. Instead, these items probably have a more local, land-based origin (beachgoers, storm-water drainage), than more heavily encrusted debris [134]. Plastic waste also affects beaches.

Plastic waste could encourage the invasion of species who prefer hard surfaces, and as a result, indigenous species may be displaced, particularly those who prefer sandy and muddy bottoms [5]. So-called "wrack" environments consist of national flotsam and jetsam, such as seaweed and driftwood that is washed up on the shore, and often contain plastic waste [5]. Beach cleanups are way to remove plastic waste, but it is often assumed that the beach will return to its previous state once the cleanup is done [5].

2.3.4 Ecotoxicology

Plastics are considered to be biochemically inert because of their macromolecular structures; they neither react with nor penetrate the cell membrane of an organism [4]. However, most plastics are not pure. Besides their polymeric structure, they consist of a variety of chemicals that all contribute to certain properties of the plastics they comprise [4]. Additives are mostly of small molecular size and are often not chemically bound to a polymer and are, therefore, able to leach from the plastics. Being primarily lipophilic, they penetrate cell membranes, interact biochemically and cause toxic effects. Moreover, plastics debris in the marine environment not only contains additives, but also contains chemicals (contaminants) adsorbed from the surrounding water [4]. The hydrophobic surface of plastics has an affinity for various hydrophobic contaminants, and these are taken up from the surrounding water and accumulate on, and in, the plastics debris. This mechanism receives great attention for microdebris or microplastics, because they are easily ingested by organisms and constitute a pathway for chemicals to enter an organism [103].

Plastics debris in the marine environment can contain two types of possible toxic contaminants, additives and hydrophobic chemicals, that become adsorbed from the surrounding water [56]. In the marine environment, absorption of contaminants by polymers is primarily studied with meso-plastic and microplastic debris. Absorption reduces the transport and diffusion of contaminants. Hydrophobic organic contaminants have a greater affinity for plastics like polyethylene, polypropylene and polyvinyl chloride (PVC) than for natural sediments [56]. Flame retirements are also present as additives in plastics and are added to many common products. The majority of flame retardants are widely used in plastics products because they affect material properties in only a minor way and are very effective in preventing ignition.

However, they are also present as contaminants almost everywhere in the world's environment; they exist in air, rivers and waters up to the Arctic regions.

Ingestion of plastic fragments by seabirds and fish may be the source of bioaccumulation of heavy metals, polychlorinated biphenyls (PCBs), dichlorodiphenyltrichloroethane (DDT) and other toxins [72]. Absorption and transfer of these chemicals by filter-feeding organisms and invertebrates may lead to reproductive disorders, disease, altered hormone levels or death at higher trophic levels [25,72,147,148].

2.4 MANAGEMENT AND PREVENTIVE MEASURES

2.4.1 Prevention and Control

Marine plastic pollution shows that we cannot really throw anything away. Reducing, reusing and recycling are the best way to stem the tide of plastics into our oceans. Moss [149] suggested the following ways in preventing and controlling plastic waste.

Hold plastic producers accountable: Many countries hold producers of materials like paints and carpets responsible for recovering and recycling their product after it is used.

Boxes should be preferred: Laundry detergent and dish soap should be in boxes instead of plastic bottles as cardboard can be more easily recycled and made into more products than plastic.

Use reusable bottles and cups: Bottled water produces 1.5 million tons of plastic waste per year, and these bottles require 47 million gallons of oil to produce. By simply refilling a reusable bottle, plastics can be prevented from ending up in the ocean.

Reduce use of plasticware: Use of silverware should be encouraged as use of plasticware will increase risk of plastic waste.

2.4.2 Management Procedures

There are various ways for management of plastic waste in the aquatic ecosystem. The following are not limited to be the only management procedures:

2.4.2.1 Public Awareness and Education

Both formal and informal education is urgently needed to raise the publics' awareness of the negative impact of irresponsible waste disposal in general and plastic waste in particular. Education must also be used to forge a positive change in attitude to plastic waste management [17]. Education is also a very powerful tool to address the issue, especially if it is discussed in schools; youngsters not only can change habits with relative ease, but also be able to take their awareness into their families and the wider community, working as catalysts for change. Since land-based sources provide major imputes of plastic debris into the oceans, if a community becomes aware of the problem and obviously is willing to act upon it, it can actually make a significant difference. The power of education should not be underestimated, and it can be more effective than strict laws, such as the Suffolk county plastics law (in New York, US) that banned some retail food packaging and was unsuccessful in reducing beach and roadside litter [150].

2.4.2.2 Recycling of Plastics through Environmentally Sound Manners

Recycling, in simple terms, is defined as the conversion of used materials (waste) into new products. The purpose of recycling is to

- Prevent waste of potentially useful materials,
- Reduce the consumption of fresh raw materials,
- Reduce energy usage, and
- Reduce air pollution (from incineration) and water pollution (from land filling) by reducing the need for "conventional" waste disposal, and lower greenhouse gas emissions as compared to virgin production [17].

According to a feasibility study on plastics waste by the Centre for Scientific and Industrial Research (CSIR) in Ghana, GHC 1,200,000 can be generated every month if plastics waste goes through various stages toward recycling. Recycling therefore provide opportunities for effective management of plastics waste management, as well as income generation. However, though some plastic wastes recycling plants have been established, the menace still persists. Funding and capacity have been identified as the major problems hindering recycling of waste. It is therefore proposed that

- A plastic waste management fund should be created to support recycling and upgrading of plastic waste infrastructure to promote private–public partnerships in the development and management of plastic waste infrastructure. The fund should be resourced with voluntary contributions from industry, government and other donors, and these contributions are tax exempt.
- Set up of differential power and water tariffs to increase the level of recycling.
- Zero-rating of recycled products to create a vibrant market for recycled products.
- Supporting the development of biodegradable bags that are more durable, reusable and recyclable.

- Preferential tax treatment to the private sector for the construction of plastics waste treatment plants.
- A voluntary code of practice for retailers, consumers and manufacturers aiming at rationalizing the issuance of plastics, increasing the usage of plastic bags made from recycled material, creation of convenient and accessible recycling stations to customers and setting up of better standards for imported packaging plastics [17].

2.4.2.3 Establishment of a Waste Stock Exchange (WSE)

One of the emerging systems that is increasingly assuming a pivotal role in the achievement of recycling and resource recovery is an online waste stock exchange (WSE). This system will serve as an online waste exchange network available for all companies in the country, to increase business profitability by promoting waste trading and recycling [151]. This system will streamline cooperation between waste producers, reusers and business advisers, making transactions quicker and easier to achieve. This web-based mechanism will serve as a secondary raw materials market that solves logistical and qualitative problems for all public and private entities that could potentially use some kind of waste in their production cycles or that implement recycling and recovery programs [151]. It is an innovative and efficient instrument, if introduced in a solid legal and economic context, and as is typical of a free and competitive market, will promote the reuse and recycling of industry by-products and wastes [151].

2.4.2.4 Conversion of Plastics Waste into Artifacts

Another option for sustainable plastic wastes management is conversion into artifacts such beads, bags, door mats and hats. This option should be promoted in basic and secondary schools [17].

2.4.2.5 Encourage the Use of Bio-Based and Biodegradable Plastics

Another way to prevent the input of persistent plastics into the marine environment is to introduce biodegradable plastics. Biodegradable plastics are made of renewable sources and consist of polymers that are capable of undergoing decomposition into carbon dioxide, water, methane, inorganic compounds or biomass. Biodegradation of these polymers is achieved by the use of microorganisms that have the ability to catabolize these polymers into less environmentally harmful materials [152]. The residue of degraded polymers is often used as plant fertilizer, and these plants can serve as a new source for manufacturing biodegradable polymers. Recently, progress has been made in developing biodegradable plastics that possess characteristic similar to those of oil-based polymers [153]. Bio-based and biodegradable plastics are on the rise, numerous new products have been developed, achievable properties are much more diverse and possible applications for these materials are much more versatile than they were just a few years ago. With regard to the end-of-life phase, biodegradable and compostable plastics offer additional recovery options, like composting or anaerobic digestion. Biodegradability is a special feature which is particularly attractive when economic and/or ecological benefits can be gained by leaving plastic products in the soil or biowaste stream. For example, used as biowaste bags, biodegradable plastics support a clean separation and collection of organic waste: divert from landfills toward high-quality compost production. Composting is important for areas where soil erosion is a serious problem [17]. Biodegradable plastics, or bioplastics, often have inferior performance compared to traditional plastics because they eventually become permeable to water. Therefore, bioplastic materials are used as disposable items, such as packaging material. The biodegradable polymers that are used are of diverse types. Bioplastics that are based on polylactic acid (PLA) and plastarch material (PSM) are two of the most commonly used ones in current commercial practice.

2.4.2.6 Impose Levies on Plastic Shopping Bags

One practical option to reduce the rate of generation of plastic waste is by discouraging overuse or misuse of plastic wrappers and carrier bags. The "abuse" of plastic shopping bags is a serious and visible environmental problem [17]. To address the "abuse" of plastic shopping bags at the source, governments should impose an environmental levy on plastic shopping bags, with the first phase covering chain and large supermarkets, convenience stores and personal health and beauty stores. Aside from reduction at the source through the environmental levy, governments should also encourage the reuse, recovery and recycling of plastic shopping bags, through source separation of waste programmer and community campaigns [17]. In some countries, whereas alternative paper wrappers/bags are free in shops, plastic wrappers/bags carry a fee that is used to subsidize the more expensive production of paper wrappers/bags. The reason is that paper waste decays, so it does not endanger the environment. However, concerns about cutting trees for paper production and its negative impact on the environment must be borne in mind. Reuse of plastic carrier bags should be vigorously encouraged and practiced by all to minimize plastic waste generation [17].

2.4.2.7 Increment in the Thickness of Plastic Films

There is also need to consider increasing the thickness of the plastic film used in manufacturing carrier bags from the current 9–11 μm to a minimum of 30 μm. This indeed is the situation in countries such as India. The increased thickness of plastic film is expected to reduce excessive contamination of plastic waste (that increases recycling costs) and makes discarded carrier bags difficult for the wind to blow around. It is proposed, however, that any increase in thickness of plastic film must be accompanied by adoption of a technology that provides a line of weakness on water sachets and other plastic containers that need to have one end torn before their contents can be used [17].

2.4.2.8 Household Segregation, Reuse and Recycling

Waste should be regarded as a great economic resource. The segregation, reuse and recycling of waste at the household level or point of generation should be encouraged. Paper, plastics, organic matter, metals and glass could all be recycled or converted to usable materials. In order to make user charges effective, there is the need to tailor such charges to the level of environmental consciousness of residents and their ability to pay [17]. Very serious consideration must be given to how more money could be generated to improve the delivery of the service. The government, in the long-term, should consider adopting and modifying the use of other economic instruments such as

- Institution of solid-waste pricing systems that will provide continuous incentives for households to reduce waste generation (e.g., pay per volume of waste).
- Disposal charges levied on dumping of industrial and municipal waste at landfill sites. Rate of charges should depend on type of waste and method of treatment before dumping.
- Incentive schemes such as subsidies, concessional loans and tax incentives to encourage governments and private investors in research, training and demonstration projects for energy resource recovery, as well as for planning for solid-waste disposal.
- Linking charges to other utility services within the district.
- Charging for dumping at landfill sites.

- Adding charges to property rates so as to unify payment.
- Indirect charges through the sale of polythene bags for waste disposal [17].

Recycled polymeric material can be reused which saves production energy and prevents the dumping of materials into the environment. During the last decade, the mechanical recycling industries have showed an encouraging trend, that is, a 7% annual growth in western Europe [2]. Unfortunately, the recycling rate varies regionally, and globally, and only a small percentage of total plastics waste is currently being recycled. In most countries, the forms of plastic that are recycled are largely limited to bottles and drink containers [14]. Most consumers are keen to recycle, and support for recycling is often very high in most Western countries. However, the differences in symbols printed in different forms of plastic to describe recyclability of the object vary considerably among countries or regions and is often an obstacle to convenient recycling. This is why, in most countries, all kinds of plastic waste are collected together and is sorted at special stations before being recycled.

2.4.2.9 Energy Recovery

Plastics are almost all derived from oil wastes which are of a high calorific value. Energy recovered from plastic waste can make a major contribution to energy production. Plastics can be co-incinerated with other wastes or used as alternative fuel (e.g., coal) in several industry processes (cement kilns). The energy content of plastic waste can be recovered in other thermal and chemical processes such as pyrolysis. As plastic waste is continuously being recycled, the plastics lose their physical and chemical properties at their end-of-life cycle. Continuous recycling could lead to substandard and low-quality products. Hence, it would no longer be economically profitable to recycle any longer [17]. Incineration with energy recovery would be the economically preferred option at this stage.

2.4.2.10 Conversion of Plastics Waste into Liquid Fuel

A research-cum-demonstration plant was set up at Nagpur, Maharashtra, for conversion of waste plastics into liquid fuel. The process adopted is based on random depolymerization of waste plastics into liquid fuel in the presence of a catalyst. The entire process is undertaken in a closed reactor vessel followed by condensation, if required. Waste plastics while heating up to 2700°C to 3000°C convert into the liquid-vapor state, which is collected in a condensation chamber in the form of liquid fuel, while the tarry liquid waste is topped-down from the heating reactor vessel. The organic gas is generated which is vented owing to lack of storage facility. However, the gas can be used in a dual fuel diesel-generator set for generation of electricity [17].

2.4.2.11 Plasma Pyrolysis Technology

In plasma pyrolysis, firstly, the plastics waste is fed into the primary chamber at 8500°C through a feeder. The waste material dissociates into carbon monoxide, hydrogen, methane, higher hydrocarbons, etc. Induced draft fan drains the pyrolysis gases as well as plastics waste into the secondary chamber, where these gases are combusted in the presence of excess air [154]. The inflammable gases are ignited with a high-voltage spark. The secondary chamber temperature is maintained at around 10,500°C. The hydrocarbon, carbon monoxide and hydrogen are combusted into safe carbon dioxide and water. The process conditions are maintained so that it eliminates the possibility of formation of toxic dioxins and furans molecules (in case of chlorinated waste) [154]. The conversion of organic waste into nontoxic gases (CO_2, H_2O) is more than 99%. The extreme conditions of plasma kill stable bacteria such as *Bacillus stereothermophilus* and *Bacillus subtilis* immediately. Segregation of the waste is not necessary, as very high temperatures ensure treatment of all types of waste without discrimination [154].

2.4.2.12 Legislative Measures

Creation of more legislative bodies such as which the United Nations Agency and the International Maritime Organization (IMO) introduced the marine pollution (MARPOL) convention in 1983, an international protocol to prevent and reduce pollution from ships. The protocol is referred to as MARPOL 73/78, from the fact that the convention was signed in 1973 and the protocol was added in 1978. The protocol has been approved by 169 countries, which together are responsible for 98% of the world's total shipping pollution, and prevents ships from releasing garbage and totally prohibits the disposal of plastics anywhere into the sea. Further, it obligates governments to keep terminal facilities and harbors clean of garbage [155]. According to the terms of this agreement, every ship having a certain weight and able to carry more than 14 persons is obligated to maintain a garbage record book, in which records of all disposal operations will be kept. Information required includes date, time, position of the ship and description and estimated amount of garbage that is incinerated or discharged. In addition to maintaining a garbage record book, marines are asked to prepare a garbage management plan that gives procedures for collecting, storing and processing on-board waste [155].

2.5 CONCLUSIONS

Because of ingestion or entanglement in plastic debris, over 270 species, including turtle, fish, seabirds, and mammals, have experienced impaired movement, starvation or death [54,156]. The important thing to do on the effect of plastics on aquatic ecosystems is to ban the input of plastics into the ocean or marine environment. Plastics do not disappear and will remain in our environments, definitely affecting wildlife, until the pollution is reduced. To avoid the death of some marine species, plastics should be prevented from entering the ocean. Strategies to ensure that materials are recycled or disposed of properly need to be developed.

Recycling is the current solution that will make plastics not to be entering oceans. Plastics contain chemicals that can be harmful to marine organisms. Education is particularly important, because it is the basis for teaching the next generation to be aware of, and address, the consequences of discarding plastics and other debris into the world oceans. With the evidence available, it is possible at present to draw sound conclusions on the direct risks posed by the occurrence of microplastics particles in the marine environment or the influence of microplastics on the risk posed to environmental and human health when associated with hazardous substances such as additives.

Plastics do not disappear and will remain in our environments indefinitely affecting aquatic life, until the pollution is reduced. Water is something every living organism on this planet cannot live without. If this resource is so precious that life cannot exist without it, we shouldn't be contaminating it [157].

The following are recommended as remedies for plastic pollution in the aquatic environment:

- Development of indigenous and homegrown technologies for plastic waste management.
- Development of quality standards and implementing them for all plastic recycled products.

- Constant information exchange on best available practices and technologies.
- Bye-laws on plastic waste management should be enforced.
- Local communities should be made aware of the health and safety aspects of all levels of plastic waste recycling.
- Policy formulation and implementation for plastic waste management should consider the needs of all levels of the community. The social responsibility for plastic waste management should be initiated at all levels of our education system, right from the primary school level upward.
- As much as possible, plastic packaging materials should be replaced with foil-coated paper.
- During waste disposal, plastic materials should be separated from other waste and treated appropriately.

REFERENCES

1. Derraik JGB. The pollution of the marine environment by plastic debris: A review. *Mar Pollut Bull* 2002, 44(9), 842–852.
2. Thompson RC, Moore CJ, VomSaal FS, and Swan SH. Plastics, the environment and human health: Current consensus and future trends. *Philos Trans R Soc Lond B Biol Sci* 2009, 364(1526), 2153–2166.
3. Wright SL, Thompson RC, and Galloway TS. The physical impacts of microplastics on marine organisms: A review. *Environ Pollut* 2013, 178(2013), 483–492.
4. Hammer J, Kraak MH, and Parsons JR. Plastics in the marine environment: The dark side of a modern gift. *Rev Environ Contam Toxicol* 2012, 220, 1–44.
5. Science for Environment Policy (SEP). Plastic waste: Ecological and human health impacts. Available online: http://ec.europa.eu/environment/integration/research/newsalert/pdf/IRI_en.pdf (accessed on 23 November 2017).
6. Koushal V, Sharma R, Sharma M, Sharma R, and Sharma V. Plastics: Issues, challenges and remediation. *Int J Waste Resour* 2014, 4(1), 134–139.
7. Plastic Ocean Foundation. A plastic ocean. Available online: http://www.plasticoceans.org (accessed on 21 November 2017).
8. United Nations Environmental Program UNEP. Marine litter - trash that kills. Available online: http://wedocs.unep.org/restbitstreams/17739/retrieve (accessed 02 January 2018).
9. United States Environmental Protection Agency (USEPA). Municipal solid waste generation, recycling and disposal in the United States: Facts and figures for 2008. Available online: http://www.paperbecause.com/PIOP/files/48/48417708-7e64/45a7-9720-d2c50e00f3b3.pdf (accessed 02 January 2018).
10. Plastics – the Fact 2014. An analysis of European plastics production, demand and waste data. Available online: http://2015.igem.org/wiki/images/e/ef/IGEM_Pasteur_Plastics_the_I (accessed on 19 January 2017).
11. UNEP. Marine litter - an analytical overview. Available online: http://www.unep.org/regionalseas/Publications/Marine_Litter.pdf (accessed on 1 December 2017).
12. Moore CJ, Leecaster MK, Moore SL, and Weisbery B. A comparison of plastic and plankton in the North plastic central Gyre. *Mar Pollut Bull* 2001, 42(12), 1297–1300.
13. Lytle GM. When the mermaids cry: The great plastic tide. Available on: http://www.plastic-pollution.org (accessed on 4 January 2018).

14. Barnes KA, Galagari F, Thompson RC, and Barlaz M. Accumulation and fragmentation of plastics debris in global environment. *Philos Trans R Soc Lond B Biol Sci* 2009, 364, 1985–1998.

15. Moore CJ. Synthetic polymers in the marine environment: A rapidly increasing, long term threat. *Environ Res* 2008, 108(2), 131–139.

16. Machuca I. Plastic pollution. Available online: http://www.oceanareachbz.com/ uploads/1/8/3/9/18397615/plastic_pollution.pdf (accessed on 12 November 2017).

17. Ampofo SK. The options for the effective management of plastic waste in Ghana. Available online: http://fonghana.org/wp-contents/uploads/2013/12/REPORT-ON-MANAGEMNT-OF-PLASTIC-WASTE-IN-GHANA-21-328-STASWAPA.pdf (accessed on 18 November 2017).

18. UNEP. Marine litter: A global challenge. Available online: http://www.unep.oer/pdf/ unep_marine_litter-a_global_challenge.pdf (accessed 04 January 2018).

19. Cooper DA. Effects of chemical and mechanical weathering processes on the degradation of plastic debris on marine beaches. PhD Thesis, The University of Western Ontario, Ontario, 2012.

20. Andrady AL. *Plastics and the Environment*. Wiley, New York, USA, 2003, 792.

21. Allsopp M, Walters A, Santillo D, and Johnston P. *Plastic Debris in the World's Ocean*. Greenpeace, Auckland, New Zealand, 2006, 44.

22. Eriksson C, and Burton H. Origins and biological accumulation of small plastic particles in fur seals from Macquarie Island. *Ambio* 2003, 32(6), 380–384.

23. Thompson RC, Olsen Y, Mitchell RP, Davis A, Rowland SJ, John AWG, McGonigle D, and Russell AE. Lost at sea: Where is all the plastic? *Science* 2004, 304, 838.

24. Gregory MR, and Ryan PG. Pelagic plastics and other seaborne persistent synthetic debris: A review of Southern Hemisphere perspectives. In: Coe JM, and Rogers DB (eds) *Marine Debris – Sources, Impacts and Solutions*. Springer-Verlag, New York, USA, 1997, pp. 49–66.

25. Ryan PG, Moore CJ, van Franeker JA, and Moloney CL. Monitoring the abundance of plastic debris in the marine environment. *Philos Trans R Soc Lond B Biol Sci* 2009, 364, 1999–2012.

26. Anonymous. *Garbage*. Center for Marine Conservation, Washington DC, USA, 1990.

27. Corbin CJ, and Singh JG. Marine debris contamination of beaches in St. Lucia and Dominica. *Mar Pollut Bull* 1993, 26, 325–328.

28. Debrot AO, Tiel AB, and Bradshaw JE. Beach debris in Curacao. *Mar Pollut Bull* 1999, 38, 795–801.

29. Galgani F, Burgeot T, Bocquene G, Vincent F, Leaute JP, Labastie J, Forest A, and Guichet R. Distribution and abundance of debris on the continental shelf of the Bay of Biscay and in Seine Bay. *Mar Pollut Bull* 1995a, 30, 58–62.

30. Galgani F, Jaunet S, Campillo A, Guenegen X, and His E. Distribution and abundance of debris on the continental shelf of the north-western Mediterranean Sea. *Mar Pollut Bull* 1995b, 30, 713–717.

31. Galgani F, Souplet A, and Cadiou Y. Accumulation of debris on the deep sea floor of the French Mediterranean Coast. *Mar Eco Prog Ser* 1996, 142, 225–234.

32. Galgani F, Leaute JP, Moguedet P, Souplets A, Verin Y, Carpenter A, Goraguer H et al. Litter on the sea floor along European coasts. *Mar Pollut Bull* 2000, 40, 516–527.

33. Garrity SD, and Levings SC. Marine debris along the Caribbean coast of Panama. *Mar Pollut Bull* 1993, 26, 317–324.

34. Gilligan MR, Randal SP, Richardson JP, and Kozel TR. Rates of accumulation of marine debris in Chatham County, Georgia. *Mar Pollut Bull* 1992, 24, 436–441.

35. Golik A. Debris in the Mediterranean Sea: Types, quantities, and behavior. In: Coe JM, and Roger DB (eds) *Marine Debris – Sources, Impacts and Solutions*. Springer-Verlag, New York, USA, 1997, pp. 7–14.

36. Hess NA, Ribic CA, and Vining I. Benthic marine debris, with an emphasis on fishery-related items, surrounding Kodiak Island, Alaska, 1994–1996. *Mar Pollut Bull* 1999, 38, 885–890.

37. Kanehiro H, Tokai T, and Matuda K. Marine litter composition and distribution on the seabed of Tokyo Bay. *Fish Eng* 1995, 31, 195–199.

38. Laist DW. Overview of the biological effects of lost and discarded plastic debris in the marine environment. *Mar Pollut Bull* 1987, 18, 319–326.

39. Lara-Dominguez AL, Villalobos-Zapata GJ, Rivera-Arriaga E, Vera-Herrera F, and Alvarez-Guillen H. Source of solid garbage in Campeche beaches, Mexico. *Revista de la Sociedad Mexicana de Historia Natural* 1994, 45, 133–142.

40. Madzena A, and Lasiak T. Spatial and temporal variations in beach litter on the Transkei coast of South Africa. *Mar Pollut Bull* 1997, 34, 900–907.

41. Manski DA, Gregg WP, Cole CA, and Richards DV. *Annual Report of the National Park Marine Debris Monitoring Program: 1990 Marine Debris Surveys*. National Park Services, Washington DC, USA, 1991.

42. Morris RJ. Floating plastic debris in the Mediterranean. *Mar Pollut Bull* 1980, 11, 125.

43. Ribic CA. Use of indicator items to monitor marine debris on a New Jersey Beach from 1991 to 1996. *Mar Pollut Bull* 1998, 36, 887–891.

44. Ribic CA, Scott WJ, and Cole CA. Distribution, type, accumulation, and source of marine debris in the United States, 1989–1993. In: Coe JM, and Roger DB (eds) *Marine Debris – Sources, Impacts and Solutions*. Springer-Verlag, New York, USA, 1997, pp. 35–47.

45. Ross SS, Parker R, and Strickland M. A survey of shoreline litter in Halifax Harbour 1989. *Mar Pollut Bull* 1991, 22, 245–248.

46. Ryan PG. The origin and fate of artefacts stranded on islands in the African sector of the Southern Ocean. *Environ Conserv* 1987, 14, 341–346.

47. Slip DJ, and Burton HR. Accumulation of fishing debris, plastic litter, and other artefacts, on Heard and Macquarie Islands in the Southern Ocean. *Environ Conserv* 1991, 18, 249–254.

48. Smith P, and Tooker J. *Marine Debris on New Zealand Coastal Beaches*. Greenpeace, Auckland, New Zealand, 1990.

49. Stefatos A, Charalampakis M, Papatheodorou G, and Ferentinos G. Marine debris on the seafloor of the Mediterranean Sea: Examples from two enclosed gulfs in Western Greece. *Mar Pollut Bull* 1999, 36, 389–393.

50. Vauk GJM, and Schrey E. Litter pollution from ships in the German Bight. *Mar Pollut Bull* 1987, 18, 316–319.

51. Walker TR, Reid K, Arnould JPY, and Croxall JP. Marine debris surveys at Bird Island, South Georgia 1990–1995. *Mar Pollut Bull* 1997, 34, 61–65.

52. Whiting SD. Types and sources of marine debris in Fog Bay, Northern Australia. *Mar Pollut Bull* 1998, 36, 904–910.

53. Williams AT, and Tudor DT. Litter burial and exhumation: Spatial and temporal distribution on a cobble pocket beach. *Mar Pollut Bull* 2001, 42, 1031–1039.

54. Laist DW. Impacts of marine debris: Entanglement of marine life in marine debris including a comprehensive list of species with entanglement and ingestion records. In: Coe JM, and Rogers DB (eds) *Marine Debris: Sources, Impacts, Solutions*. Springer-Verlag, New York, USA, 1997, pp. 99–139.

55. Aliani S, and Molcard A. Hiteh-hiking on floating marine debris: Macrobiotic special in the Western Mediterranean Sea. *Hydrobiol* 2003, 503(11), 59–61.

56. Teuten EL, Saquing JM, Knappe DRU, Barlaz MA, Jonsson S, Bjorn A, Rowland SJ et al. Transport and release of chemicals from plastics to the environment and to wildlife. *Philos Trans R Soc Lond B Biol Sci* 2009, 364(1526), 2027–2045.

57. Secretariat of the Convention on Biological Diversity and the Scientific and Technical Advisory Panel - Global Environment Policy (GEF). Impacts of Marine Debris on Biodiversity: Current Status and Potential Solutions. CBD Technical Series No. 67, Montreal, 2012. Available online: http://www.cbd.int/doc/publications/cbd-ts-67-en (accessed on 27 December 2017).

58. Goldberg ED. Plasticizing the sea floor: An overview. *Environ Technol* 1997, 18, 195–202.

59. Gregory MR. Environmental implications of plastic debris in marine settings entanglement, ingestion, smothering, hangers-on, hitch-hiking and alien invasions. *Philos Trans R Soc Lond B Biol Sci* 2009, 364, 2013–2025.

60. Quayle DV. Plastics in the marine environment: Problems and solutions. *Chem Ecol* 1992, 6, 69–78.

61. Wilber RJ. Plastics in the North Atlantic. *Oceanus* 1987, 30, 61–68.

62. Mcllgorm A, Campbell HF, and Rule MJ. The economic cost and control or marine debris damage in the Asia-plastic region. *Ocean Coastal Manage* 2011, 54(9), 643–651.

63. Wolfe DA. Persistent plastics and debris in the ocean: An international problem of ocean disposal. *Mar Pollut Bull* 1987, 18, 303–305.

64. Stamper MA, Spicer CW, Neiffer DL, Mathews KS, and Fleming GJ. Morbidity in a juvenile green sea turtle (Chelonian mydas) due to ocean-borne plastic. *J Zoo Wildl Med* 2009, 40(1), 196–198.

65. Schrey E, and Vauk GJM. Record of entangled Gannets(SualBassana) at Aelgoland, German Bright. *Mar Pollut Bull* 1987, 18(6), 350–352.

66. Moser ML, and Lee DS. A Fourteen-year survey of plastic ingestion by Western North Atlantic Seabirds. *Colonial Water Birds* 1992, 15, 83–94.

67. Robards MD, Piah JF, and Wohl KD. Increasing frequency of plastic particles ingested by seabirds in the subarctic North Pacific. *Mar Pollut Bull* 1995, 30, 151–157.

68. Blight LK, and Burger AE. Occurrence of plastics particles in seabirds from the Eastern North pacific. *Mar Pollut Bull* 1997, 34, 323–325.

69. Azzarello MY, and Van-Vleet ES. Marine birds and birds and plastic pollution. *Mar Ecol Prog Ser* 1987, 37, 295–303.

70. Carpenter EJ, Anderson SJ, Harvey GR, Miklas HP, and Peck BB. Polystyrene spherules in coastal waters. *Science* 1972, 178, 749–750.

71. Rothstein SI. Plastic particle pollution of the surface of the Atlantic Ocean: Evidence from a seabird. *Condor* 1973, 75, 344–345.

72. Ryan PF. Effects of ingested plastic on seabird feeding: Evidence from chickens. *Mar Pollut Bull* 1988, 19, 125–128.

73. Ziko V, and Hanlon M. Another source of pollution by plastics skin cleaners with plastic scrubbers. *Mar Pollut Bull* 1991, 22, 41–42.

74. Fry DM, Fefer SI, and Sileo L. Ingestion of plastic debris by laysan albatross and wedge-tailed shearwaters in the Hamwaiian islands. *Mar Pollut Bull.* 1987, 180, 339–343.

75. Provencher JF, Gaston AJ, Mallocy ML, O'Hara PD, and Gilcgirst HG. Ingested plastic in a diving seabird, the thick-billed Murre (Uria lomvia) in the Eastern Canadian Arctic. *Mar Pollut Bull* 2010, 60(9), 1406–1411.

76. Avery-Gomm S, Prevencher JF, Morgan KH, and Bertram DF. Plastic ingestion in marine-associated bid species from the Eastern North pacific. *Mar Pollut Bull* 2013, 72(1), 257–259.
77. Van Franeker JA, Blaize CJ, Danielsen J, Fairclough K, Gollan J, Guse N, and Turner DM. Monitoring plastic ingestion by the northern fulmar Fulmarus glacialis in the North Sea. *Environ Pollut* 2011, 159(10), 2609–2615.
78. Boerger CM, Lattin GL, Moore SL, and Moore CJ. Plastic ingestion by planktivorous fish in the North Pacific Central Gyre. *Mar Pollut Bull* 2010, 60(12), 2275–2278.
79. Plot V, and Georges JY. Plastic debris in a nesting leatherback turtle in French Guiana. *Chelonian Conserv Biol* 2010, 9(2), 267–270.
80. Sancho G, Puente E, Bilbao A, Gomez E, and Arregi L. Catch rates of monkfish (Lophius spp.) by lost tangle nets in the Cantabrian Sea (northern Spain). *Fish Res* 2003, 64, 129–139.
81. Bullimore BA, Newman PB, Kaiser MJ, Gilbert SE, and Lock KM. A study of catches in a fleet of "ghost fishing" pots. *Fish Bull* 2001, 99, 247–253.
82. Joint Nature Conservation Committee (JNCC). Ghost fishing. Available online: http://www.jncc.gov.uk/page-1567 (accessed 02 December 2017).
83. Baird RW, and Hooker SK. Ingestion of plastics and unusual prey by a juvenile harbour porpoise. *Mar Pollut Bull* 2000, 40(8), 719–720.
84. Jones MM. Fishing debris in the Australian marine environment. *Mar Pollut Bull* 1995, 30, 25–33.
85. Feldkamp S, Costa D, and Dekrey GK. Energetic and behavioural effects of net entanglement on juvenile northern fur seals Callorhinus ursinus. *Fish Bull* 1989, 87, 85–94.
86. Secchi ER, and Zarxur S. Plastic debris ingested by a blain Ville's beaked whale, Mesoplodon densirostris washed ashore in Brazil. *Aquat Mammal* 1999, 25(1), 21–24.
87. Jacobsen JK, Massey L, and Gullan F. Fatal ingestion of floating net debris by two sperm whales (Physeter macrocephalus). *Mar Pollut Bull* 2010, 60, 765–767.
88. Ayalon O, Goldrath T, Rosethal G, and Grossman M. Reduction of plastics carrier bag use: An analysis alternative in Israel. *Waste Manage* 2009, 29, 2025–2032
89. Simmonds MP. Cetaceans and marine debris: Great unknown. *J Mar Biol* 2012, 2012, 1–8.
90. De Stephanis R, Gimenez J, Carpinelli E, Gutierrez-Exposito C, and Canadas A. As main meal for sperm whales: Plastics debris. *Mar Pollut Bull* 2013, 69(1), 206–214.
91. Baulch S, and Perry C. A sea of plastic: Evaluating the impacts of marine debris on cetaceans. *Mar Pollut Bull* 2014, 80(1), 210–221.
92. Allen R, Jarvis D, Sayer S, and Mills C. Entanglement of grey seals Halicho ersgrypus at a hail out site in Cornwll, UK. *Mar Pollut Bull* 2012, 64(12), 2815–2819.
93. Mattlin RH, and Cawthorn MW. Marine debris - an international problem. *N Z Environ* 1986, 51, 3–6.
94. Gramentz D. Involvement of loggerhead turtle with the plastic metal and hydrocarbon pollution in the central Mediterranean. *Mar Pollut Bull* 1988, 19, 11–13.
95. Bugoni L, Krause I, and Etry MV. *Mar Pollut Bull* 2001, 42, 1330–1334.
96. Duguy R, Moriniere P, and Lemilianire C. Factors of mortality of marine turtles in the Bay of Biscay. *Oceanol Acta* 1998, 21, 383–388.
97. Carr A. Impact of nondegradable marine debris on the ecology and survival outlook of sea turtles. *Mar Pollut Bull* 1987, 18, 352–356.
98. Balazs G. Impact of ocean debris on marine turtles: Entanglement and ingestion. In: Shomura, RS, and Yoshida HO (eds) *Proceedings of the Workshop on the Fate*

and Impact of Marine Debris, Honolulu, 27–29 November 1984, U.S. Department of Commerce, NOAA Technical Memorandum NMFS SWFC-54, USA, 1985, 387–429.

99. O'Hara K, Ludicello S, and Bierce R. *A Citizen's Guide to Plastics in the Ocean: More than a Litter Problem*. Centre for Marine Conservation, Washington DC, 1988, 142.

100. Ward JE, and Kach DJ. Marine aggregates facilitate ingestion of nanoparticles by suspension-feeding bivalves. *Mar Environ Res* 2009, 68(3), 137–142.

101. Goldstein MC, Rosenberg M, and Cheng L. Increased oceanic microplastic debris enhances oviposition in an endemic pelagic insect. *Biol Lett* 2012, 8(5), 817–820.

102. Graham ER, and Thompson JT. Deposit – and suspension-feeding sea cucumbers (Echinodermata) ingest plastic fragments. *J Exp Mar Biol Ecol* 2009, 368(1), 22–29.

103. Andrady AL. Microplastics in the marine environment. *Mar Pollut Bull* 2011, 62(8), 1596–1605.

104. Beck CA, and Barros NB. The impact of debris on the Florida manatee. *Mar Pollut Bull* 1991, 22, 508–510.

105. Campagna C, Falabella V, and Lewis M. Entanglement of southern elephant seals in squid fishing gear. *Mar Mamm Sci* 2007, 23, 414–418.

106. Croxall JP, Rodwell S, and Boyd IL. Entanglement in man-made debris of Antarctic fur seals at Bird Island, South Georgia. *Mar Mamm Sci* 1990, 6, 221–233.

107. Dau BK, Gilardi KVK, Gulland FM, Higgins A, Holcomb JB, St. Leger J, and Ziccardi MH. Fishing gear-related injury in California marine wildlife. *J Wildl Dis* 2009, 45(2), 355–362.

108. Fowler CW. Marine debris and northern fur seals: A case study. *Mar Pollut Bull* 1987, 18, 326–335.

109. Good TP, June JA, Etnier MA, and Broadhurst G. Derelict fishing nets in Puget Sound and the Northwest Straits: Patterns and threats to marine fauna. *Mar Pollut Bull* 2010, 60, 39–50.

110. Moore E, Lyday S, Roletto J, Litle K, Parrish JK, Nevins H, Harvey J et al. Entanglements of marine mammals and seabirds in central California and the north-west coast of the United States 2001–2005. *Mar Pollut Bull* 2009, 58, 1045–1051.

111. Pham CK, Gomes-Pereria JN, Isidro EJ, Santos RS, and Morato T. Abundance of litter on Condor seamount (Azores, Portugal, Northeast Atlantic). *Deep-Sea Res II* 2013, 98, 204–208.

112. Velez-Rubio GM, Estrades A, Fallabrino A, and Tomas J. Marine turtle threats in Uruguayan waters: Insights from 12 years of stranding data. *Mar Biol* 2013, 160, 2797–2811.

113. Winn JP, Woodward BL, Moore MJ, Peterson ML, and Riley JG. Modeling whale entanglement injuries: An experimental study of tissue compliance, line tension, and draw-length. *Mar Mamm Sci* 2008, 24, 326–340.

114. Woodward BL, Winn JP, Moore MJ, and Peterson ML. Experimental modeling of large whale entanglement injuries. *Mar Mamm Sci* 2006, 22, 299–310.

115. Bjorndal KA, Bolten AB, and Lagueux CJ. Ingestion of marine debris by juvenile sea turtles in coastal Florida habitats. *Mar Pollut Bull* 1994, 28, 154–158.

116. Brandão ML, Braga KM, and Luque JL. Marine debris ingestion by Magellanic penguins, Spheniscus magellanicus (Aves: Sphenisciformes), from the Brazilian coastal zone. *Mar Pollut Bull* 2011, 62, 2246–2249.

117. Browne MA, Niven SJ, Galloway TS, Rowland SJ, and Thompson RC. Microplastic moves pollutants and additives to worms, reducing functions linked to health and biodiversity. *Curr Biol* 2013, 23, 2388–2392.

118. Carey MJ. Intergenerational transfer of plastic debris by short-tailed shearwaters (Ardenna tenuirostris). *Emu* 2011, 111, 229–234.
119. Cedervall T, Hansson LA, Lard M, Frohm B, and Linse S. Food chain transport of nanoparticles affects behaviour and fat metabolism in fish. *PLOS ONE* 2012, 7, e32254.
120. Connors PG, and Smith KG. Oceanic plastic particle pollution: Suspected effect on fat deposition in red phalaropes. *Mar Pollut Bull* 1982, 13, 18–20.
121. Lee KW, Shim WJ, Kwon OY, and Kang JH. Size-dependent effects of micro polystyrene particles in the marine copepod Tigriopus japonicus. *Environ Sci Technol* 2013, 47, 11278–11283.
122. Oliveira M, Ribeiro A, Hylland K, and Guilhermino L. Single and combined effects of microplastics and pyrene on juveniles (0+ group) of the common goby Pomatoschistus microps (Teleostei, Gobiidae). *Ecol Indic* 2013, 34, 641–647.
123. Rochman CM, Browne MA, Halpern BS, Hentschel BT, Hoh E, Karapanagioti HK, Rios-Mendoza LM, Takada H, The S, and Thompson RC. Policy: Classify plastic waste as hazardous. *Nature* 2013a, 494, 169–171.
124. Rochman CM, Hoh E, Hentschel B, and Kaye S. Long-term field measurement of sorption of organic contaminants to five types of plastic pellets: Implications for plastic marine debris. *Environ Sci Technol* 2013b, 47, 1646–1654.
125. Rochman CM, Hoh E, Kurobe T, and Teh SJ. Ingested plastic transfers hazardous chemicals to fish and induces hepatic stress. *Sci Rep* 2013c, 3, 3263.
126. von Moos N, Burkhardt-Holm P, and Kohler A. Uptake and effects of microplastics on cells and tissue of the blue mussel Mytilus edulis L. after an experimental exposure. *Environ Sci Technol* 2012, 46, 11327–11335.
127. Katsanevakis S, Verriopoulos G, Nicolaidou A, and Thessalou-Legaki M. Effect of marine litter on the benthic megafauna of coastal soft bottoms: A manipulative field experiment. *Mar Pollut Bull* 2007, 54, 771–778.
128. Lewis CF, Slade SL, Maxwell KE, and Matthews TR. Lobster trap impact on coral reefs: Effects of wind driven trap movement. *N Z J Mar Freshw Res* 2009, 43, 271–282.
129. Uneputty P, and Evans SM. The impact of plastic debris on the biota of tidal flats in Ambon Bay (Eastern Indonesia). *Mar Environ Res* 1997, 44, 233–242.
130. Chiappone M, Dienes H, Swanson DW, and Miller SL. Impacts of lost fishing gear on coral reef sessile invertebrates in the Florida Keys National Marine Sanctuary. *Biol Conserv* 2005, 121, 221–230.
131. Chiappone M, White A, Swanson DW, and Miller SL. Occurrence and biological impacts of fishing gear and other marine debris in the Florida Keys. *Mar Pollut Bull* 2002, 44, 597–604.
132. Uhrin AV, and Schellinger J. Marine debris impacts to a tidal fringing-marsh in North Carolina. *Mar Pollut Bull* 2011, 62, 2605–2610.
133. Özdilek HG, Yalcin-Özdilek S, Ozaner FS, and Sönmez B. Impact of accumulated beach litter on Chelonia mydas L. 1758 (green turtle) hatchlings of the Samandağ Coast, Hatay, Turkey. *Fresen Environ Bull* 2006, 15, 95–103.
134. Widmer WM, and Hennemann MC. Marine debris in the island of Santa Catarina, South Brazil: Spatial patterns, composition, and biological aspects. *J Coastal Res* 2010, 26: 993–1000.
135. Rochman CM, Browne MA, Underwood AJ, van Franeker JA, Thompson RC, and Amaral-Zettler LA. The ecological impacts of marine debris: Unraveling the demonstrated evidence from what is perceived. *Ecology* 2016, 97, 302–312.

136. Law KL. Plastics in the marine environment. *Annu Rev Mar Sci* 2017, 9, 205–229.
137. Ryan PG, and Swanepoel D. Cleaning beaches: Sweeping the rubbish under the carpet. *S Afr J Sci* 1996, 92, 163–165.
138. Ballance A, Tyan PG, and Turpie JK. How much is a clean beach worth? The impact of litter on beach users in the Cape Peninsula, South Africa. *S Afr J Sci* 2000, 96, 210–213.
139. Costa MF, Ivar do Sul JA, Silva-Cavalcanti JS, Araujo MCB, Spengler A, and Tourinho PS. On the importance of size of plastic fragments and pellets on the strandline: A snapshot of a Brazilian beach. *Environ Monit Assess* 2010, 168, 299–304.
140. Winston JE, Gregory MR, and Stevens LM. Encrusters, epibionts and other biota associated with pelagic plastics: A review of biogeographical, environmental, and conservation issues. In Coe JM, and Rogers DB (eds) *Marine Debris, Sources, Impacts, and Solutions.* Springer-Verlag, New York, USA, 1997, pp. 81–97.
141. Gregory MR, and Andrady AL. Plastics in the marine environment. In Andrady AL (ed) *Plastics and the Environment.* Wiley, New Jersey, USA, 2003, pp. 379–401.
142. McKinney RL. On predicting biotic homogenization - species-area patterns in marine biota. *Global Ecol Biogeogr Lett* 1998, 7, 297–301.
143. Barnes DKA. Invasions by marine life on plastic debris. *Nature* 2002, 416, 808–809.
144. Winston JE. Drift plastic – an expanding niche for a marine invertebrate? *Mar Pollut Bull* 1982, 13, 348–357.
145. Gregory MR. The hazards of persistent marine pollution: Drift plastics and conservation islands. *J R Soc N Z* 1991, 21, 83–100.
146. Gregory MR. Plastics and the South Pacific Island shores: Environmental implications. *Ocean Coastal Manage* 1999, 42, 603–615.
147. Gregory MR. Plastic "scrubbers" in hand cleanser: A further (and minor) source of marine pollution identified. *Mar Pollut Bull* 1996, 32(12), 867–811.
148. Lee K, Tanabe S, and Koh C. Contamination of polychlorinated biphenyls (PCB's) in sediments from Kyeonggi Bay and nearby areas, Korea. *Mar Pollut Bull* 2001, 42, 273–279.
149. Moss L. 16 simple ways to reduce plastic waste. Available online: http://www.mnn.com/lifestyle/responsible-living/stories/16-simple-ways-reduce-plastic-waste (accessed on 17 November 2017).
150. Ross SS, and Swanson RL. The impact of the Suffouk County, New York plastics ban on beach and roadside litter. *J Environ Syst* 1995, 23, 337–351.
151. Coda Canati F. Secondary raw materials market creation: Waste stock exchange. FEEM Working Paper No. 28.2000. 2000. Available online at http://dx.doi.org/10.2139/ssrn.229277 (accessed on 20 November 2017).
152. Bioplastics 24. How plastics are made from plants. Available online: http://www:bioplasatics24.com/index.php (accessed on 12 November 2017).
153. Song JH, Murphy RJ, Narayan R, and Davies GBH. Biodegrable and compostable Alternatives to conventional plastics. *Philos Trans R Soc Lond B Biol Sci* 2009, 364(1526), 2127–2139.
154. Kumar GM, Irshad A, Raghunath BV, and Rajarajan G. Waste management in food packaging industry. In Prashanthi M, and Sundaram R (eds) *Integrated Waste Management in India: Status and Future Prospects for Environmental Sustainability.* Springer, Berlin, Germany, 2016, Vol. 10, pp. 265–277. ISBN: 978-3-319-27226-9.
155. International Maritime Organization (IMO). International convention for the prevention of pollution from ships 1973, as modified by the protocol of 1978 relating thereto (MARPOL). 2010. Available online: http://www/imo.org/conventions.contents.asp?doc-id=678andtopic-id=258#garbage (accessed on 12 November 2017).

156. Wabnitz C, and Nichols WJ. Plastic pollution: An ocean emergency. *Mar Turtle Newsl* 2010, 129, 1–4.

157. Blackwell B. Record levels of plastic pollution found in Lake Erie, Northeast Ohio Media Group. Available online: http://www.cleveland.com/metro/index.ssf/2012/12/record_levels_of_plastic_poll ut.html (accessed on 12 November 2017).

SECTION II

Plastics in the Environment

CHAPTER 3

Microplastics and Nanoplastics in the Environment

M. Humam Zaim Faruqi and Faisal Zia Siddiqui

CONTENTS

3.1 INTRODUCTION

The term "plastic" is derived from the ancient Greek word *plastikos*, which refers to something that is appropriate for molding, and the Latin term *plasticus*, which means molding or shaping. The advent of plastics has led to fundamental transformation in the quality of human life and has facilitated unprecedented technological advancements. Because of its use in diverse sectors such as construction, packaging, transportation, healthcare and electronics, the development of this versatile and ubiquitous class of materials is widely perceived as one of the greatest technological achievements of the 21st century. Plastics are composed of large chain-like macromolecules which further consist of many recurring smaller molecules connected in a sequence. A substance with this kind of molecular arrangement is known as a "polymer." Every individual molecule in a polymer chain is a single unit and is known as a "monomer." Thus, monomers are small molecules that possess the ability to bond together to form long chains. A vast majority of polymers in today's world are synthetic plastics. However, many natural polymers such as deoxyribonucleic acid (DNA) also exist. Therefore, it is important to note that although all plastics are polymers, not all polymers are plastics [1].

The global production of plastics increased 20-fold from 15 million tonnes (MT) in 1964 to about 335 MT in 2016 [2,67]. The production is further expected to double again in the next 20 years and quadruple by 2050 [2]. Asia produced nearly half of the world's plastic in 2014. China and Japan, with 26% and 4% of the global plastic production, respectively, are two of the highest plastic producers in Asia [3]. About 50% of the produced plastics are utilized for single-use disposable applications, such as packaging, agricultural films and consumer items; 20%–25% for long-term infrastructure such as pipes, cable coatings and structural materials and the remainder for durable consumer applications such as electronic goods, furniture, vehicles, etc. [8]. The increasing consumption of plastics can be attributed to the low cost, ease of handling and good durability of plastic material. Unfortunately, the growing demand of plastic products to advance our lifestyle has trapped us into a vicious cycle of producing, consuming and disposing them. The huge plastic waste generated worldwide indicates their widespread use in the urban settlements. However, the quantities of plastics that are reused or recycled are significantly low. The global rate of recycling of plastics across developed countries is about 17%, which is far lower than that of paper (58%) and iron and steel (70%–90%) [1,4,5]. In 2012, only about 6 MT (26%) and 2.5 MT (9%) of post-consumer plastics were recycled in the European Union (EU) and US, respectively. However, more than 8.75 MT of plastic in the EU and 29 MT in the US was discarded without recycling. The likely reason for this is the increase in manufacturing costs by about 20% when using recycled plastics, as opposed to utilizing virgin-plastic feedstock [1]. Further, the repeated melting and molding of a normal commodity plastic also results in diminished mechanical properties and reduced durability.

Other than the plastic waste which is recycled, most of the plastic ends up in landfills, where it takes a few hundred years to decompose [6]. However, a significant proportion of plastics end up unaccounted for in the environment and get deposited in terrestrial and marine ecosystems. While the dumping of plastics in terrestrial ecosystems occurs mainly from littering or illegal land disposal, plastic items end up as marine debris owing to littering, insufficient treatment capacity, transport by wind or surface runoff, illegal discharges of domestic and industrial wastewaters, accidental inputs and coastal human activities [8,11]. In terrestrial ecosystems, the ingestion of plastics by livestock may lead to nutritional deficiencies in the animals. Plastic pollution has attracted attention as a growing threat to marine ecosystems. Conservative estimates of overall financial damage of plastics to marine ecosystems stood at $13 billion per year in 2014 [12]. Marine environments consists of about 10% of the total plastics produced [7]. It is estimated that of all the waste present in the aquatic environment, about 60%–80% is plastic [10]. As plastic is a synthetic product, its sources are mostly inland [11]. The quantity of plastic waste available to enter the marine environment from land is alarming. It is estimated to increase manifold with insufficient waste management infrastructure improvements [9]. A study conducted by United Nations Joint Group of Experts on the Scientific Aspects of Marine Pollution (GESAMP) concluded that 80% of the waste in the marine environment originates from land, while only 20% was a result of activities at sea [1].

The larger plastic debris, known as "macroplastics" tend to accumulate in specific areas of the ocean as a result of convergence of surface currents [11]. As the plastic waste enters into the marine environment, various factors begin to have a degrading effect on the plastic material. Long-chain plastic macromolecules are degraded into shorter chains of lower molecular weight. Plastic degradation can occur under the influence of various environmental factors such as heat, light, mechanical force, water and chemicals [1,12]. The degradation can be classified as thermal, photooxidative, ozone-induced, catalytic or mechanochemical degradation or biodegradation [13]. Prolonged exposure to UV radiation from sunlight can cause oxidation of the polymer matrix, resulting in bond cleavage [6]. Such degradation may result in leaching out of additives, which are originally meant to enhance durability and corrosion resistance of plastics [22]. Because of the loss of structural integrity, these plastics become susceptible to fragmentation resulting from abrasion, turbulence and wave action [19,23]. Consequently, the macroplastics degrade over spatial and temporal variations into smaller pieces. As the production of plastic has increased exponentially over the last five decades, the smaller pieces have progressively become more widespread in the marine environment. These small plastic particles can be classified based on their size as "microplastics" or "nanoplastics" [1].

Microplastics were first introduced to the world in 1972 when small-sized plastic particles were reported to be floating on the surface of the Sargasso Sea [14]. Microplastic was defined by the Steering Committee of National Oceanic and Atmospheric Administration (NOAA) Marine Debris Program as being a plastic less than 5 mm in size along its longest dimension [15]. Microplastic generally refers to any piece of plastic smaller than 5 mm to 1 μm in size along its longest dimension. A piece of plastic less than 1 μm in size is referred to as a nanoplastic. Due to the extremely small size of nanoplastics and the associated difficulties in their detection and recovery, most studies on the marine ecosystems tend to overlook nanoplastics and instead focus on microplastics. However, the presence of nanoplastics in the marine environment is likely to become increasingly significant in the years to come, and researchers have already begun to speculate on the effect of nanoplastics on the marine food web [6].

3.2 SOURCES OF MICROPLASTICS

A wide variety of polymers are used to make a range of plastic products. Almost all of the common polymers have been recognized as microplastic particles. These include polyamide (PA) (nylon), polyethylene (PE), polypropylene (PP), polystyrene (PS), polyvinyl alcohol (PVA), polyvinyl chloride (PVC), polyethylene terephthalate (PET), etc. However, a single polymer can be used in a variety of applications. Therefore, any information on the type of polymer found as a microplastic particle may not be particularly beneficial in identifying the source of the particle [16]. Microplastics can be divided into two major categories depending upon their potential sources and usage: primary and secondary microplastics.

3.2.1 Primary Microplastics

Primary microplastics are the particles that are manufactured for use as microplastic particles (<5 mm in size). The release of primary microplastics into the environment may occur because of the spillage of pre-production pellets (~4 mm in diameter) or powders (>1 mm in diameter). Primary microplastics are also used as shot-blasting media during the cleaning of softer metals like aluminum and for removing paint or rust from machinery, engines and boat hulls [19,20,21]. The use of exfoliant microbeads (typically 250 μm in diameter) as microplastic scrubbers in cosmetic products like exfoliant hand cleansers and facial scrubs also comprise primary sources of microplastics in the environment [6,16].

3.2.2 Secondary Microplastics

Microplastics that result from environmental fragmentation of larger items of plastic debris such as packaging, ropes and fishing nets are termed secondary microplastics. A combination of physical, chemical or biological processes can reduce the structural integrity of plastic debris over time and lead to fragmentation [19]. The quantiles of secondary microplastics broadly reflect the quantities of larger, identifiable items of plastic debris collected during routine monitoring [16]. A substantial input of fibers from textiles has been found in residues from sewage treatment plants on land [17] and at former sewage sludge dumping grounds in the marine environment [18]. Such fibers may enter the environment as particles that are already microplastic in nature. However, since these were not manufactured as microplastics, they are generally classified as secondary microplastics.

3.3 MICROPLASTICS SAMPLING AND ABUNDANCE

The distribution of microplastics can be considered from several perspectives [16]. The most apparent is perhaps the geographic distribution, which may be evaluated on a range of spatial scales, ranging from a few meters within a beach [26] to several kilometers between countries or continents [18,27]. The distribution of microplastics can also be categorized across environmental compartments, such as the quantity of microplastics along coastlines, at the sea surface, in the water column or on the sea bed (both subtidal and intertidal), and quantities in biota that have been accumulated by ingestion [16,31]. Temporal distribution, although less popular, can be used to assess the variation in marine debris over time and the effect of legislation to curb the release of microplastics into the environment. Smaller plastic debris (<0.5 mm in diameter) is considered a widely under-researched component of marine debris due to the difficulties in assessing the density, abundance, and distribution of this component within the marine environment [6,24]. A quantification of the input of plastics in a marine environment is limited by the following factors [6,16,24,25]:

- The wide array of pathways by which plastics may enter the marine environment such as rivers, surface runoff, municipal and industrial wastewater treatment, etc.
- The requirement of accurate timescales for which plastics remain in the marine bodies prior to degradation
- The complexities arising due to the vastness of the oceans compared to the size of plastics being assessed
- The spatial and temporal variability caused by oceanic currents and seasonal variations
- The variation in sampling methods due to the absence of universally recognized protocols for collection or sampling

3.3.1 Sampling Techniques

Despite the difficulties posed by the limitations described above, a variety of sampling techniques have been developed that allow the determination of spatial and temporal distribution of plastic debris. These include beach combing, sediment sampling, marine trawls, marine observational surveys and biological sampling [6].

3.3.1.1 Beach Combing

Beach combing is a sampling technique that is typically used by researchers and environmental awareness groups, and involves the collection and identification of plastic litter items, in a systematic manner along a specified stretch of coastline. Beach combing at regular intervals allows the monitoring of plastic debris accumulation over time [6,25]. However, microplastics invisible to the naked eye tend to go unnoticed by using such a technique [6]. Further, plastic debris along a coastline consists of a mix of terrestrial litter left by recreational beach activities and sea-deposited debris, and may not be an accurate indicator of plastic debris in the marine environment [28].

3.3.1.2 Sediment Sampling

Sediment sampling involves the assessment of benthic material from estuaries, beaches and the seafloor for the presence of microplastics. Several methods that are used for sediment sampling include the use of saline water or mineral sediments for low-density microplastic flotation, application of lipophilic dye to stain the plastic particles and assist in microscopic techniques, and the use of Fourier-transform infrared (FTIR) spectra to compare with known polymers [23,30,31].

3.3.1.3 Marine Trawls

The microplastic samples within a water column can be collected by conducting marine trawls using fine meshes. Different types of trawls that can be used for sampling include mantra trawls for sampling surface water, bongo nets for mid-water level sampling and benthic trawls for assessing seabed [25,26,31]. The presence of microplastics can then be determined by examination of the collected samples under a microscope or investigating the residue left behind after allowing the seawater to evaporate [6,30].

3.3.1.4 Marine Observational Surveys

The marine surveys involve the assessment of the type, size and location of visible plastic debris by observers on boats or other sea vessels [6]. While the surveys are helpful in determining macroplastic debris over larger areas, the detection of microplastics using this technique is difficult. Further, since the debris is not collected, no further assessment of litter can be done. The subjective nature of observations during these surveys increases the susceptibility to bias [25,32].

3.3.1.5 Biological Sampling

Several marine animals can mistake plastic fragments for prey and consume them. The dissection of marine animals washed ashore or prompting regurgitation in seabirds allows the analysis of their gut contents for the presence of plastic, which can then be identified and quantified [33,34,35].

The microplastics sampled from marine environments are typically extracted by density separation, filtered, and are purified by rinsing, chemical or enzymatic digestion. Smaller plastic particles require visualization via microscopy and formal identification using pyrolysis gas chromatography (pyrolysis-GC) or FTIR/Raman spectroscopy, which require expensive capital investments and are often time-consuming. Further, the characterization of plastic particles into different shapes, sizes and colors and their subsequent quantification play important roles in the accurate estimation of microplastics abundance.

3.3.2 Spatial and Temporal Variation of Microplastics

Oceanographic modeling indicates that a significant proportion of floating debris reaching the oceans is accumulated in "gyres" [6]. Subtropical gyres are large-scale systems of wind-driven surface currents which flow clockwise in the northern hemisphere and counterclockwise in the south. This flow is caused by the Coriolis effect, a force which tends to move wind and water currents to the right in the northern hemisphere and to the left in the south, thus creating cyclonic movements in the atmosphere and oceans [16]. The earliest attempts to mathematically reproduce the probable pathways of marine debris were conducted using a global set of trajectories of satellite-tracked drifters. A probabilistic model eliminates the bias in spatial distribution of drifter data due to heterogenous deployments [16,46]. A study conducted for over 6000 plankton tows undertaken between 1986 and 2008 in the North Atlantic Ocean and the Caribbean Sea showed that plastic constituted 60% of the samples [36]. In this study, distinct spatial patterns of plastic were found. About 83% of total sampled plastic was found in subtropical latitudes, with the highest concentration mapped to the North Atlantic gyre ($20,328 \pm 2,324$ pieces/km^2) [6,36]. An even higher plastic concentration of 33,271 pieces/km^2 has been found in the North Pacific gyre [37]. The findings of such large quantities have led to the North Pacific gyre being labeled as "plastic soup" and described as the "great Pacific garbage patch" [38].

Plastic particles found in the marine environment are composed of a variety of polymers. Depending upon their density, composition and shape, the plastic particles can be fully buoyant (that is, float on the surface), be neutrally buoyant or sink to the seabed [6]. Density plays an important role in the location of plastic particles within the water column in marine bodies. Low-density microplastics have been predominantly found on the sea-surface layer [20,21]. However, a study involving the collection of microplastics from the North Atlantic found microplastics with a density greater than that of water to be floating on the surface waters [39]. The average density of seawater is 1025 kg/m^3 [41]. In another study, 90% of the microplastics recovered from the western North Atlantic Ocean had an average density less than the density of seawater, with densities ranging from 808 to 1238 kg/m^3 [40]. The occurrence of higher density microplastics on the surface of oceans may result from the powerful upward and downward movements of water due to temperature difference at different depths or the presence of air bubbles or pockets in denser microplastics which tend to increase their buoyancy [1]. However, PVC (with a density of 1150–1700 kg/m^3) and polyamide (with a density of 1120–1380 kg/m^3) may transport to various marine regions as a result of wind and tidal currents, rather than due to density variation. Similarly, some microplastics, like polyethylene, which possess a density of about 920–970 kg/m^3, undergo an increase in density as a result of weathering. Several studies have shown that the growth of fouling organisms like biomass can sink the buoyant microplastics [20,26,31,42]. Plastic debris in marine bodies rapidly accumulate microbial biofilms, which allow the colonization of algae and invertebrates on the plastic surface, thus increasing the density of plastic particles [30].

High-density microplastics that are composed of materials like polyester, PVC and polyamide are found in largest quantities in the benthic zones. The determination of microplastic quantities on the seabed is constrained by the cost and difficulties in sampling [42]. Some countries such as the Netherlands and Scotland have initiated "fishing for litter" schemes and submersible video recordings to document the microplastic quantity on the seafloor [6,43,44]. However, microplastics may eventually escape the lower detection limits of these sampling methods [6]. When high-density microplastics enter the seas, they tend to remain in suspension owing to high flow rate, tidal fronts or a large surface area [26]. These dense microplastics tend to sink to the seabed when the momentum is lost [42].

The slow emergence of global legislation governing the indiscriminate disposal of plastics has led to an increase in microplastics entering the marine bodies, corresponding to the higher production rates of plastic [25,42,45]. While progressive fragmentation over time and the popularity of plastic scrubbers have increased the volume of microplastic debris in the ocean, it has caused a reduction in the average size of plastic litter over time [6,42]. During the analysis of samples from the North Sea and Northwest Atlantic, microplastics concentration in the 1980s and 1990s were found to be substantially higher than those in the 1960s and 1970s [31].

3.4 FATE OF MICROPLASTICS AND NANOPLASTICS IN THE ENVIRONMENT

3.4.1 Tracking the "Missing" Plastic

Plastic debris is abundant and extensively spread throughout the marine habitat. Plastic pollution in marine environments has been recognized as a global environmental threat [45,47]. Microplastics have amassed on the shorelines from the poles to the equator, at the sea surface and in the deep sea and in terrestrial habitats [16]. However, one of the crucial mysteries in the field of microplastic and nanoplastic research is that based on the amount of plastics entering the ocean, 99% of the plastics that should be present in the oceans is missing and cannot be accounted for [1]. The surface load and size distribution of floating plastic debris has been found to be well below that expected from the generation of plastics and their input rates in oceans [49]. However, no definite temporal trend in floating plastic concentration has been reported, perhaps because of confounded spatial and temporal variability in data [49,50]. A study carried out in 1999 discovered that the smaller the size of a microplastic particle, the lower was its abundance in the North Pacific Central gyre [45]. These observations support the premise of substantial loss of plastic from the ocean surface. The following possible sinks may be proposed as the eventual fate of micro- and nano-sized plastics [49].

3.4.1.1 Nanofragmentation

The gap in the distribution of plastic size below 1 mm could indicate a rapid breakdown of the plastic fragments from the millimeter to the micrometer scale. In some recent studies, the scanning electron micrographs of microplastic surfaces have indicated that in addition to solar-induced fragmentation, oceanic bacterial populations may also contribute to the degradation of plastics on the surface of oceans [51]. Limited information on physical, chemical and biological processes driving plastic disintegration suggest that two-phase fragmentation, with accelerated breakdown of photo-degraded fragments, may also occur. Nanofragmentation can render the smaller plastic pieces undetectable to conventional sampling nets. The abundance of nano-scale plastic particles is still not quantified in the open ocean [30,49].

3.4.1.2 Biofouling

Small-sized plastic particles, with high surface-to-volume ratios, may preferentially submerge in the ocean waters by ballasting due to epiphytic growth. Biofouled plastic fragments are often incorporated in the sediment in shallow, nutrient-rich areas. However, this mechanism may not be applicable in the deep, open ocean [30,52]. Defouling of plastic in deep waters may occur from adverse conditions for epiphytic organisms, causing the plastic particle to return to the surface [49].

3.4.1.3 Marine Ingestion

Zooplanktivorous organisms like epipelagic and mesopelagic fish are an abundant trophic assemblage in the ocean, and accidental ingestion of plastic is known to occur during their feeding activity [49]. The most frequent plastic size ingested by these fish has been reported to range between 0.5 and 5.0 mm, matching the predominant size of plastic debris which experience global losses in oceans [53,54,55]. Small mesopelagic fish represent the most pervasive zooplanktivorous collection in the open ocean, with densities close to one individual per square meter, even in the oligotrophic subtropical gyres [49,56]. The plastic fragments ingested by smaller fish can be transferred to larger predators, sink with the bodies of dead fish or be defecated [49,54,55].

3.4.2 Interaction of Chemical Pollutants with Microplastics

The industrial, agricultural and other anthropological activities on land lead to significant pollution of the aquatic environment with hazardous chemicals [1,57]. Approximately 625,000 barrels of oils are discharged into the coastal waters every year as a consequence of runoff from land. Plastic materials have been estimated to release between 35 and 917 tonnes of chemical additives into the marine environment every year, with a majority being released from plasticized PVC [1,58].

Plastic fragments which get deposited in the shallow or deep waters of the marine environment not only pose physical pollution problems but are also a potential source of significant chemical hazard. Although plastics are considered to be biochemically inert material which do not interact with the human endocrine system owing to their large molecular size, the plastic debris existing in the marine environment may well carry chemicals of smaller molecular size which can enter the cell, chemically interact with important biological entities or molecules and disrupt the endocrine system [48]. Such chemicals can be categorized as [1,48]:

- Chemical pollutants, which can get absorbed into the bulk of plastic material by diffusion
- Hydrophobic chemicals, which can be adsorbed from the surrounding seawater owing to their affinity for the hydrophobic surface of plastics
- Additives, monomers and oligomers of the component molecules of the plastics

Many of these chemical pollutants are considered to be persistent because they are highly resistant to environmental degradation and consequently remain in the marine environment for considerable periods of time [1,61]. In a study conducted for contaminants of growing concern, the time period in which a contaminant decreases from being of highest concern to a baseline level of the lowest concern was typically 14.5 ± 4.5 years [59]. Many such pollutants can bioaccumulate in organisms, including humans, and possess the ability to traverse the food chain as well as being passed from mothers to their offspring [60]. The persistent hazardous chemicals of this category are termed as persistent organic pollutants (POPs). Some classes of pollutants have been demonstrated to interact with microplastics and are designated as POPs. The Stockholm Convention on Persistent Organic Pollutants was created in 2001 by the United Nations Environment Programme (UNEP) with an objective to protect human health and environment from POPs. The convention lists POPs with considerable health effects and undertakes measures to reduce, restrict or eliminate the manufacture of POPs [1]. The chemicals that are included in the list follow specific criteria, such as possessing a half-life of at least two months in the aquatic environment, showing evidence of widespread dispersal from its source, remaining stable and demonstrating bioaccumulative and toxicological effects [63].

The list initially consisted of 12 chemical pollutants, namely aldrin, chlordane, dichlorodiphenyltrichloroethane (DDT), dieldrin, endrin, hexachlorobenzene, heptachlor, mirex, polychlorinated biphenyls, polychlorinated dibenzofurans (PCDF), polychlorinated dibenzo-p-dioxins (PCDD) and toxaphene that were known as the "dirty dozen." It has now been extended to include up to 26 chemical substances of potential concern. Some other classes of pollutants that pose considerable hazards to biological organisms include perfluoroalkylates, phthalates, polybrominated diphenyl ethers (PBDEs), polychlorinated biphenyls (PCBs), and polycyclic aromatic hydrocarbons (PAHs). The plastic monomer bisphenol A (BPA) and alkylphenol additives have been known to exert estrogenic effects while some phthalate plasticizers are associated with reduced testosterone production [48]. PCBs and PBDEs are persistent, bioaccumulative and toxic (PBT) chemicals and pose serious concern to human health which has resulted in their inclusion in the List of chemicals for Priority Action adopted by OSPAR (Oslo and Paris Conventions) Commission [61,62]. PCBs are highly resilient in nature and have been estimated to remain as the most widespread contaminant in the aquatic environment and organisms until at least 2050 [64].

The interaction of microplastics with POPs involves three distinct phenomena [1]:

- *Absorption*: Waterborne chemical pollutants diffuse into the bulk of the plastic material by absorption. The PAH phenanthrene diffuses into the bulk of polypropylene matrix by surface diffusion.
- *Adsorption*: During the adsorption of POPs to the microplastics, the POP acts as the adsorbate and microplastic as the adsorbent/substrate. Some microplastics, such as polyethylene, polystyrene and polypropylene are nonpolar in nature, while some others, such as polyamide and polycarbonate, are polar. The more nonpolar a plastic material is, the greater is its affinity for hydrophobic POPs.
- *Desorption*: The pollutants which contaminate microplastics may be desorbed after ingestion of the microplastic by aquatic organisms. A study has demonstrated that the microplastics composed of polyethylene that are contaminated with the PAH phenanthrene exhibited the highest potential for transfer from microplastics to organisms [66].

Microplastics in the marine environment can be capable of concentrating waterborne POPs by up to 1 million times higher than their background concentration in the seawater [65]. The ingestion of contaminated microplastics by aquatic organisms represents a unique pathway for introduction and subsequent traversing of toxic chemicals into the food web. However, there is wide scope for research in the ways by which microplastics can act as vectors for the introduction of chemicals in the marine food web.

3.5 THE WAY FORWARD

Macroplastics are progressively fragmenting into smaller microplastics and nanoplastics which cannot be captured in routine monitoring studies. Ultraviolet light weakens plastic, and coupled with biological (bacterial fragmentation) or mechanical (wave energy) action, this can cause large items to fragment into micro- and nanoplastics [49]. Further, it has been found that microplastics have accumulated in inaccessible and relatively under-sampled locations such as the deep sea, within the arctic ice and biota [16]. Microplastics have also accumulated in beach sediments more than a meter underneath the sediment surface [29].

It seems probable that the eventual fate of all plastic in the environment is as microplastic- or nanoplastic-sized fragments. In some locations, microplastics are substantially more profuse in quantity than macroplastics. However, macroplastics are still the dominant size fraction by mass in the environment. Therefore, even if additional input of plastic debris into the oceans is prevented forthwith, the quantity of micro- and nanoplastics will continue to increase over time owing to fragmentation of legacy plastic items already present in the environment [16]. It is also established that there are no effective resources for eliminating microplastics or nanoplastics once they are in the ocean. Hence, in addition to the quantification of microplastic and nanoplastic abundance and the consideration of their potential health and environmental effects, it is imperative to focus on developing methods to reduce the input of plastic debris into the ocean. Some countries have effected important changes in their legislations pertaining to plastic waste management and disposal in order to achieve the required cut in plastic input to oceans. For example, India amended its Plastic Waste Management Rules in 2018 to phase out all multilayered plastic (MLP) which is non-recyclable, non-energy recoverable or with no alternate use. Efforts are also being undertaken by the Indian government to enforce a complete ban on single-use plastics in the near future.

It is usually considered that with the exception of incinerated plastic, all the conventional (non-biodegradable) plastic that has ever been produced is still present on the planet Earth in a form that is too large in size to be biodegraded. Further, the rate of biodegradation of plastics is very slow, to the extent that biodegradation cannot be relied upon to have any useful effect on the vast quantities of plastic debris in the oceans [16].

The above facts point to an urgent need for undertaking concrete steps toward prevention of indiscriminate and irresponsible use of plastics around the world. Some important considerations that need to be made relating to the input of plastic debris into the marine environment are:

- About 8% of the world's oil production is utilized to produce plastic items [16]. However, 50% of these items are discarded within a short time frame [8]. Since the plastic products are inherently recyclable, it is possible to decrease the accumulation of plastic debris as well as the demand for fossil carbon by recycling end-of-life plastics.
- A notable difference between the problem of marine plastic debris and several other existing environmental issues is that the emission of plastic debris to the oceans is not directly linked to the benefit. The benefits from plastics can be obtained without there being a need for end-of-life emissions of plastic to the marine environment [16].
- While the relative importance of certain impacts caused by marine debris is debatable, there is typically universal consensus among scientists, industry representatives, policy makers and non-governmental organizations (NGOs) that the unchecked input of marine debris is detrimental to the economy, wildlife and the environment [16].

The solutions to the plastic menace are well-known and lie on land rather than at the sea. The problems that retard progress toward sustainable use of plastics relate to prioritizing the solutions at hand. Some potential solutions to ensure judicious use and minimal disposal of plastics are described as follows:

- Reduction in the usage of plastic products leads to reduction in the amount of new plastic produced and ultimately reduces the potential for micro- or nanoplastic formation.
- Reusing the plastic items directly reduces the need for new plastic items, thereby also reducing the quantity of end-of-life plastic material.
- Recycling the end-of-life plastic material back into new items in a closed loop can simultaneously reduce the accumulation of waste and the demand for fossil carbon.
- Recovering the energy potential of plastics by incineration helps in extraction of the value of plastic material that cannot be reused or recycled.

REFERENCES

1. Crawford CB, and Quinn B. *Microplastic Pollutants*. n.p.: Elsevier Inc, 2017.
2. Neufeld L, Stassen F, Sheppard R, and Gilman T (eds) *The New Plastics Economy: Rethinking the Future of Plastics (Industry Agenda)*. n.p.: World Economic Forum, 2016.
3. Wu C, Zhang K, and Xiong X. Microplastic pollution in Inland waters focusing on Asia. In: Wagner M, and Lambert S (eds) *Freshwater Microplastics: Emerging Environmental Contaminants, The Handbook of Environmental Chemistry 58*. Springer Nature, Cham, 2018, pp. 85–99.
4. International Council of Forest and Paper Associations. *Statement on Paper Recycling*. International Council of Forest and Paper Recycling, Washington, 2015.
5. United Nations Environment Programme (UNEP). *Recycling Rates of Metals: A Status Report, A Report of the Working Group on the Global Metal Flows to the International Resource Panel*. n.p.: United Nations Environment Programme, 2011.
6. Cole M, Lindeque P, Halsband C, and Galloway TS. Microplastics as contaminants in the marine environment: A review. *Mar Pollut Bull*. 2011, 62(12), 2588–2597.
7. Thompson RC. Plastic debris in the marine environment: Consequences and solutions. *Mar Nat Conserv Eur*, 2006, 193, 107–115.
8. Hopewell J, Dvorak R, and Kosior E. Plastics recycling: Challenges and opportunities. *Philosophical Trans The Royal Soc B Biol Sci* 2009, 364, 2115–2126.
9. Jambeck JR, Geyer R, Wilcox C, Siegler TR, Perryman M, Andrady A, Narayan R et al. Plastic waste inputs from land into the ocean. *Science*. 2015, 347(6223), 768–770.
10. Bejgarn S, MacLeod M, Bogdal C, and Breitholtz M. Toxicity of leachate from weathering plastics: An exploratory screening study with *Nitocra spinipes*. *Chemosphere*. 2015, 132, 114–119.
11. Rocha-Santos TAP, and Duarte AC (eds). *Characterization and Analysis of Microplastics*. n.p. Elsevier, 2017.
12. Wang J et al. The behaviors of microplastics in the marine environment. *Marine Environ Res* 2016, 113, 7–17.
13. Singh B, and Sharma N. Mechanistic implications of plastic degradation. *Polymer Degrad Stab* 2008, 93, 561–584.
14. Carpenter EJ, and Smith KL Jr. Plastics on the Sargasso Sea surface. *Science*. 1972, 175(4027), 1240–1241.

15. Arthur C, Baker J, and Bamford H (eds). Proceedings of the International Research Workshop on the Occurrence, Effects and Fate of Microplastic Marine Debris. *National Oceanic and Atmospheric Administration Technical Memorandum NOS-OR&R-30 Silver Spring*, NOAA Marine Debris Division, MD, 2009.

16. Thompson RC. Sources, distribution, and fate of microscopic plastics in marine environments. In: Takada H, and Karapanagioti HK (eds) *Hazardous Chemicals Associated with Plastics in the Marine Environment, The Handbook of Environmental Chemistry 78*, Springer Nature, Cham, 2019, pp. 121–133.

17. Habib D, Locke DC, and Cannone LJ. Synthetic fibers as indicators of municipal sewage sludge, sludge products, and sewage treatment plant effluents. *Water, Air and Soil Pollution* 1998, 103, 1–8.

18. Browne MA et al. Accumulation of microplastic on shorelines worldwide: Sources and sinks. *Environ Sci Technol.* 2011, 45(21), 9175–9179.

19. Browne MA, Galloway T, and Thompson RC. Microplastic—an emerging contaminant of potential concern? *Integr Environ Assess Manag.* 2007, 3(4), 559–561.

20. Derraik JGB. The pollution of the marine environment by plastic debris: A review. *Mar Pollut Bull.* 2002, 44, 842–852.

21. Gregory MR. Plastic 'Scrubbers' in hand cleansers: A further (and minor) source for marine pollution identified. *Mar Pollut Bull.* 1996, 32(12), 867–871.

22. Talsness CE, Andrade AJM, Kuriyama SN, Taylor JA, and vom Saal FS. Components of plastic: Experimental studies in animals and relevance for human health. *Philos Trans R Soc Lond B Biol Sci.* 2009, 364, 2079–2096.

23. Barnes DKA, Walters A, and Gonçalves L. Macroplastics at sea around Antarctica. *Mar Environ Res.* 2010, 70, 250–252.

24. Doyle MJ, Watson W, Bowlin NM, and Sheavly SB. Plastic particles in coastal pelagic ecosystems of the Northeast Pacific ocean. *Mar Environ Res.* 2011, 71, 41–52.

25. Ryan PG, Moore CJ, van Franeker JA, and Moloney CL. Monitoring the abundance of plastic debris in the marine environment. *Philos Trans R Soc Lond B Biol Sci.* 2009, 364, 1999–2012.

26. Browne MA, Galloway TS, and Thompson RC. Spatial patterns of plastic debris along estuarine shorelines. *Environ Sci Technol.* 2010, 44(9), 3404–3409.

27. Collignon A et al. Neustonic microplastic and zooplankton in the North Western Mediterranean Sea. *Mar Pollut Bull.* 2012, 64, 861–864.

28. OSPAR Commission. *Monitoring of Marine Litter in the OSPAR, OSPAR Pilot Project on Monitoring Marine Beach Litter.* n.p.: OSPAR Commission, 2007.

29. Claessens M, De Meester S, van Landuyt L, De Clerck K, and Janssen CR. Occurrence and distribution of microplastics in marine sediments along the Belgian coast. *Mar Pollut Bull.* 2011, 62, 2199–2204.

30. Andrady AL. Microplastics in the marine environment. *Mar Pollut Bull.* 2011, 62, 1596–1605.

31. Thompson RC et al. Lost at sea: Where is all the plastic?. *Science.* 2004, 304, 838.

32. Pruter AT. Sources, quantities and distribution of persistent plastics in the marine environment. *Mar Pollut Bull.* 1987, 18(68), 305–310.

33. Blight LK, and Burger AE. Occurrence of plastic particles in seabirds from the Eastern North Pacific. *Mar Pollut Bull.* 1997, 34(5), 323–325.

34. van Franeker JA. *Fulmar Litter EcoQO Monitoring in the Netherlands 1979–2008 in relation to EU directive 2000/59/EC on Port Reception Facilities, IMARES Report CO27/10.* Institute for Marine Resources & Ecosystem Studies Wageningen UR, Amsterdam, 2010.

35. Tourinho PS, Ivar do Sul JA, and Fillmann G. Is marine debris ingestion still a problem for the coastal marine biota of southern Brazil?. *Mar Pollut Bull* 2010, 60, 396–401.

36. Law KL et al. Plastic accumulation in the North Atlantic subtropical Gyre. *Science.* 2010, 329(5996), 1185–1188.

37. Moore CJ et al. A comparison of plastic and plankton in the North Pacific Central Gyre. *Mar Pollut Bull.* 2001, 42(12), 1297–1300.

38. Kaiser J. The dirt on ocean garbage patches. *Science.* 2010, 328(5985), 1506.

39. Colton JB Jr., Knapp FD, and Burns BR. Plastic particles in surface waters of the Northwestern Atlantic. *Science.* 1974, 185, 491–497.

40. Moret-Ferguson S et al. The size, mass, and composition of plastic debris in the western North Atlantic Ocean. *Mar Pollut Bull.* 2010, 60, 1873–1878.

41. Rios LM et al. Quantitation of persistent organic pollutants adsorbed on plastic debris from the North pacific Gyre's "eastern garbage patch". *J Environ Monit.* 2010, 12, 2226–2236.

42. Barnes DKA et al. Accumulation and fragmentation of plastic debris in global environments. *Philos Trans R Soc Lond B Biol Sci.* 2009, 364, 1985–1998.

43. Lozano RL, and Mouat J (eds) *Marine Litter in the North-East Atlantic Region: Assessment and Priorities for Response.* OSPAR Commission, London, 2009.

44. Watters DL et al. Assessing marine debris in deep seafloor habitats off California. *Mar Pollut Bull.* 2010, 60, 131–138.

45. Moore CJ. Synthetic polymers in the marine environment: A rapidly increasing, long-term threat. *Environ Res.* 2008, 108, 131–139.

46. Maximenko N. Pathways of marine debris derived from trajectories of Lagrangian drifters. *Mar Pollut Bull.* 2012, 65, 51–62.

47. Baztan J et al. (eds). *MICRO 2016. Fate and Impact of Microplastics in Marine Ecosystems: From the Coastline to the Open Sea.* n.p.: Elsevier, 2017.

48. Teuten EL et al. Transport and release of chemicals from plastics to the environment and to wildlife. *Philos Trans R Soc Lond B Biol Sci.* 2009, 364, 2027–2045.

49. Cozar A et al. Plastic debris in the open ocean. *Proc the Mem Natl Acad Sci the USA* 2014, 111(28), 10239–10244.

50. Law KL et al. Distribution of surface plastic debris in the Eastern Pacific Ocean from an 11-year data set. *Environ Sci Technol.* 2014, 48, 4732–4738.

51. Zettler ER, Mincer TJ, and Amaral-Zettler LA. Life in the "Plastisphere": Microbial communities on plastic Marine Debris. *Environ Sci Technol.* 2013, 47, 7137–7146.

52. Vianello A. Microplastic particles in sediments of Lagoon of Venice, Italy: First observations on occurrence, spatial patterns and identification. *Estuarine, Coast Shelf Sci* 2013, 130, 54–61.

53. Lusher AL, McHugh M, and Thompson RC. Occurrence of microplastics in the gastrointestinal tract of pelagic and demersal from the English Channel. *Mar Pollut Bull.* 2013, 67, 94–99.

54. Foekema EM et al. Plastic in North Sea fish. *Environ Sci Technol.* 2013, 47, 8818–8824.

55. Eriksson C, and Burton H. Origins and biological accumulation of small plastic particles in fur seals from Macquarie Island. *Ambio: A J the Human Environment* 2003, 32(6), 380–384.

56. Lam VWY, and Pauly D. Mapping the global biomass of mesopelagic fishes. *Sea Around US* 2005, 30, 4.

57. Webster L et al. An assessment of persistent organic pollutants (POPs) in wild and rope grown blue mussels (Mytilius edulis) from Scottish coastal waters. *J Environ Monit.* 2009, 11, 1169–1184.

58. Suhrhoff TJ, and Scholz-Bottcher BM. Qualitative impact of salinity, UV radiation and turbulence on leaching of organic plastic additives from four common plastics: A lab experiment. *Mar Pollut Bull.* 2016, 102(1), 84–94.

59. Halden RU. Epistemology of contaminants of emerging concern and literature meta-analysis. *J Hazard Mater.* 2015, 282, 2–9.

60. Kalantzi OI et al. Polybrominated diphenyl ethers and selected organochlorine chemicals in grey seals (Halichoerus grypus) in the North Sea. *Chemosphere.* 2005, 58, 345–354.

61. Webster L et al. Halogenated persistent organic pollutants in relation to trophic level in deep sea fish. *Mar Pollut Bull.* 2014, 88, 14–27.

62. OSPAR Commission. *OSPAR List of Chemicals for Priority Action (Reference No. 2004-12), OSPAR Convention for the Protection of the Marine Environment of the North-East Atlantic.* n.p.: OSPAR Commission, 2013.

63. Harrison RM (ed). *Pollution: Causes, Effects, and Control* (5th ed.). Royal Society of Chemistry Publishing, Cambridge, 2014.

64. Tanabe S. PCB problems in the future: Foresight from current knowledge. *Environ Pollut.* 1988, 50, 5–28.

65. Wright SL, Thompson RC, and Galloway TS. The physical impacts of microplastics on marine organisms: A review. *Environ Pollut.* 2013, 178, 483–492.

66. Bakir A, Rowland SJ, and Thompson RC. Transport of persistent organic pollutants by microplastics in estuarine conditions. *Estuarine, Coas Shelf Sci* 2014, 140, 14–21.

67. Hamid FS et al. Worldwide distribution and abundance of microplastic: How dire is the situation? *Waste Manag Res.* 2018, 36(10), 873–897.

CHAPTER 4

Marine Pollution
*The Problem of Microplastics**

A. D. O. Santos, M. T. L. Nascimento, M. de Oliveira e Sá,
D. M. Bila, J. A. Baptista Neto, R. Pereira and M. N. Vieira

Throughout the history of mankind and the conquest of the seas and oceans, these water masses were always seen as unlimited sinks of wastes since they were assumed as being able to disperse, dilute and redistribute natural and synthetic substances. However, in the last few decades, we have finally realized that this capacity is not unlimited [1,2]. According to the literature, each year about 6.4 million tons of litter are deposited in oceans and seas. Per year, there are about 8 million tons of daily sewage, together with 5 million tons of solid residues, thrown into the marine environment by boats. Moreover, it was estimated that more than 13,000 plastic pieces are floating per each square kilometer of ocean [3]. The marine litter is a great and crescent environmental threat since it can be found in all oceans and seas, even in remote places far away from obvious sources of pollution. Marine litter can migrate long distances through oceans currents and winds being observed in marine and coastal environments, from poles to equator, from continental littorals to small remote islands. Islands completely made of litter already exist and the slow degradation process of litter aggravates this problem [4].

The concern about the presence of plastics in marine environments comes from many years ago. Actually, in 1972, Carpenter et al. [5] warned about the increase of plastic production, which could lead to greater concentrations of plastics on the sea surface. Only a few months later, it was reported the first case of plastics ingestion by fishes [5]. Nowadays, it is known that several million tons of plastics have been produced ever since [6,7,8], leading to the need to deal with this contamination, especially in oceans where plastics suffer degradation and fragmentation [6,8]. Their main sources are beach litter (contributing to about 80% of plastic debris), the fishing industry (about 18%) and aquaculture [8,9]. Coastal tourism, recreational and commercial fishing and marine vessels may also be the origin of plastic pollution [10]. Plastics debris migrates through

* Previously published in *J Marine Sci Res Dev* 5(3): 167. doi:10.4172/2155-9910.1000167. © 2015 Santos ADO et al. This is an open-access article distributed under the terms of the Creative Commons Attribution License, which permits unrestricted use, distribution, and reproduction in any medium, provided the original author and source are credited.

the oceans being transformed into small fragments forming microplastics. Microplastics receive this designation owing to their size smaller than 5 mm [11].

The impacts of microplastics still remains unclear; nevertheless, some conclusions and suspicions were already raised by recent studies. In fact, the evidence of exposure of several marine organisms is large, although it is difficult to quantify such exposures and to establish dose/effect relationships required for setting risk limits (as PNEC values predicted no effect concentrations), as we intended to demonstrate. Table 4.1 summarizes some of the studies that were performed, mainly aimed in detecting exposures to microplastics and in evaluating resulting effects on different species, both under natural and laboratorial conditions. Some review papers also summarize much more information analyzing data with different points of view [12–14]. However, all the existing data suggest that for assessing the risks of microplastics, dose–response curves have to be established under laboratorial conditions, and for being representative of field situations, such doses for different species have likely to be established, at least based on the size, concentration of particles and on the chemical composition of the microplastics. These seems to be the most relevant factors determining their bioavailability, chemical versus physical effects and potential for transference through trophic chains [12].

Nevertheless, the contamination of marine environments by microplastics may have other types of impacts, indirectly affecting organisms. The ingestion of microplastics by small animals may cause a decrease in food consumption due to satiation feeling and/or intestinal blockage leading to death [15]. These compounds can accumulate in the gut of filter-feeding mussels, persisting for more than 48 days [16].

It is known, for example, that the presence of small plastics debris in beach sand slows the heating of the sediments [17]. The resulting reduction in temperature of sand can impact organisms with temperature-depending sex determination, like turtles that can be affected even by a low concentration of plastic (1.5%) [18]. Further, the permeability of the sand increases with the presence of microplastics [17]. This change combined with grain size increase and desiccation stress could negatively affect the embryonic development of eggs of several organisms like crustaceans [19], mollusks [20], polychaetes [21] and fishes [22]. Permeability increase also leads to a change in trace element cycling in beach sediments. When sands have a higher permeability, more water is flushed through the beaches, giving more oxygen and organic matter to the small interstitial organisms. More oxygen and organic matter causes an increase in the abundance of such organisms, which, in turn, will release a higher amount of compounds resulting from their metabolism in water. More metabolites lead to changes in oxygen gradients and redox conditions impacting those environments [17,23].

Besides all the efforts applied at regional, national and international levels, marine litter continues to increase. Delays in the application and fulfillment of the already-existent regulations, or either the lack of supervision or of specific regulations in several parts of the world, are contributing to such an increasing problem. More awareness and outreaching activities to the general public are also required to promote new behaviors related with plastics use and disposal. Such actions are of particular importance since the effects of marine pollution with microplastics are still not evident for society, thus resulting, for example, in misinterpretations of the taxes applied to plastic bags. Nevertheless, there is still hope that, as it happened with other dangerous contaminants such as tributyltin [23], the legislation will contribute to prevent the catastrophe presently envisaged to the marine environment due to pollution with microplastics.

TABLE 4.1 Collection of Some Microplastics Exposure and Effects in Animal Species under Both Natural and Laboratorial Conditions

Local	Species	Goals	Main Results	References
Laboratorial exposure	*Lytechinus variegatus* (sea urchin)	Compare the effects of plastic pellets (virgin and beach stranded) on *Lytechinus variegatus* embryo development	A 58.1% and 66.5% increase of anomalies in embryonic development were recorded for beach stranded and virgin pellets, respectively. The pellets were tested in a proportion of 1:4 (pellet:seawater).	[24]
Laboratorial exposure	*Mytilus edulis* (mussel)	Assess the uptake and translocation of microplastics (10–30–90 mm) under laboratorial conditions and the effects on energy metabolism	Organisms exposed to a high concentration of polystyrene microspheres (110 particles/mL sea water). Microplastics were present in all organisms collected (0.2 ± 0.3 particles/g body weight). Ingestion and translocation of microplastics in the gut did not affect the cellular energy allocation.	[25]
Laboratorial exposure		Assess effects of polyethylene ingestion at cellular and subcellular levels	After intake of particles with 0–80 μm, the following effects were observed: strong inflammatory response; granulocytoma formation after lysosomal membrane destabilization in connective tissue of digestive gland. Microplastic uptake into the gills and stomach with transport to digestive gland where they accumulated in lysosomal system in 3 h.	[26]
Laboratorial exposure		Assess the effects of 30 nm polystyrene particles (0, 0.1, 0.2, and 0.3 g/L) on the feeding behavior	Filtering activity was reduced in presence of polystyrene. Production of pseudo-feces when exposed to 0.1 g/L. The polystyrene was recognized as a low nutritional food by mussels.	[27]

(Continued)

TABLE 4.1 (*Continued*) Collection of Some Microplastics Exposure and Effects in Animal Species under Both Natural and Laboratorial Conditions

Local	Species	Goals	Main Results	References
Laboratorial exposure		Evaluation of the ingestion, translocation and accumulation of microplastics debris (3.0 or 9.6 μm)	Microplastics accumulation in gut. Microplastics capture in hemolymph. Microplastics translocation from gut to circulatory system during 48 days.	[28]
Laboratorial exposure		Assess the presence of microplastics in soft tissues (whole body except the shell)	0.36 ± 0.07 particles/g (wet weight).	[29]
Laboratorial exposure	*Mytilus galloprovincialis* (mussel)	Evaluate the effects of pyrene in presence of polyethylene and polystyrene microplastics	Presence of microplastics in hemolymph, gills and in digestive glands. Microplastics caused DNA strand breaks in hemocytes at 20 g/L. Pyrene effects were emphaticized by microplastics because they adsorb pyrene, increasing its uptake and bioavailability.	[30]
Laboratorial exposure	*Crassostrea gigas* (oyster)	Assess the presence of microplastics in soft tissues (whole body except the shell)	0.47 ± 0.16 particles/g (wet weight).	[29]
Laboratorial exposure	*Arenicola marina* (annelid)	Assess the uptake and translocation of microplastics (10–30–90 mm) under laboratorial conditions and the effects on energy metabolism	Organisms exposed to a high concentration of polystyrene microspheres (110 particles/g sediment). Microplastics were present in all organisms collected in the field; on average 1.2 ± 2.8 particles/g body weight. Ingestion and translocation of microplastics in the gut did not affect the cellular energy allocation.	[25]

(*Continued*)

TABLE 4.1 (*Continued*) Collection of Some Microplastics Exposure and Effects in Animal Species under Both Natural and Laboratorial Conditions

Local	Species	Goals	Main Results	References
Laboratorial exposure	*Arenicola marina* (annelid)	Assess the bioaccumulation of polystyrene and polychlorinated biphenyl	A low polystyrene dose increased bioaccumulation of PCBs by a factor of 1.1–3.6. Polystyrene did not accumulate in *A. marina*, but it can be ingested by its predators while in the gut of *A. marina*.	[31]
Laboratorial exposure		Evaluation of the effects of microscopic unplasticized polyvinylchloride (UPVC)	Energy reserves depletion after a chronic exposure to a dose of UPVC corresponding to 5% of sediment weight. Accumulation of UPVC in longer gut and inflammation with an enhanced phagocytic response after a chronic exposure.	[32]
Laboratorial exposure	*Pomatoschistus microps* (common goby fish)	Assess the predatory behavior of juveniles in the presence of microplastics	Microplastics (420–500 μm size) were ingested, suggesting confusion with food. Such confusion was dependent on the color of the microplastics and on the conditions of the fish juveniles.	[33]
Laboratorial exposure	*Pomatoschistus microps* (common goby fish)	Assess the Influence of microplastics on chromium toxicity in juveniles	In presence of microplastics (0.216 mg/L), chromium (1.8–28.4 mg/L) inhibited acetylcholinesterase activity.	[34]
Northwestern Mediterranean basin	Zooplankton	Evaluation of the ratio of microplastic to zooplankton in neustonic waters collected in 40 sampling stations	Presence of microplastics of different types (filaments, polystyrene, thin plastic films) in 90% of the sampling stations, with sizes ranging 0.3–0.5 mm and an average weight of 1.81 mg/particle.	[35]

(*Continued*)

TABLE 4.1 (*Continued*) Collection of Some Microplastics Exposure and Effects in Animal Species under Both Natural and Laboratorial Conditions

Local	Species	Goals	Main Results	References
			A ratio of 1:5 (microplastic:zooplankton) was recorded in neustonic water samples, thus representing a high risk to filter-feeding organisms.	
Southwest of Plymouth, United Kingdom	*Cepola macrophthalma* (bandfish)	Assessment of plastic ingestion (The study documents microplastics in 10 species of fish from the English Channel.)	Microplastics ingestion (<40 pieces/particles). Presence of polyamide, semi-synthetic cellulosic material and rayon in gastrointestinal tracts).	[36]
Southwest of Plymouth, United Kingdom	*Callionymus lyra* (common dragonet fish)	Assessment of plastic ingestion (The study documents microplastics in 10 species of fish from the English Channel.)	Microplastics ingestion (<50 pieces/particles). Presence of polyamide, semi-synthetic cellulosic material and rayon in gastrointestinal tracts).	[36]
Southwest of Plymouth, United Kingdom	*Buglossisium luteum* (yellow sole)	Assessment of plastic ingestion (The study documents microplastics in 10 species of fish from the English Channel.)	Microplastics ingestion (<20 pieces/particles). Presence of polyamide, semi-synthetic cellulosic material and rayon in gastrointestinal tracts).	[36]
Southwest of Plymouth, United Kingdom	*Microchirus variegatus* (sole)	Assessment of plastic ingestion (The study documents microplastics in 10 species of fish from the English Channel.)	Microplastics ingestion (<20 pieces/particles). Presence of polyamide, semi-synthetic cellulosic material and rayon in gastrointestinal tracts).	[36]
Southwest of Plymouth, United Kingdom	*Aspitrigla cuculus* (red gurnard fish)	Assessment of plastics ingestion (The study documents microplastics in 10 species of fish from the English Channel.)	Microplastics ingestion (<70 pieces). Presence of polyamide, semi-synthetic cellulosic material and rayon in gastrointestinal tracts).	[36]

(*Continued*)

TABLE 4.1 (*Continued*) Collection of Some Microplastics Exposure and Effects in Animal Species under Both Natural and Laboratorial Conditions

Local	Species	Goals	Main Results	References
Mediterranean Sea (Pelagos Sanctuary)	*Balaenoptera physalus* (fin whale)	Detection of microplastics and phthalates in surface neustonic/planktonic samples; detection of phthalates in stranded fin whales	56% of the surface neustonic/planktonic samples contained microplastic particles. Portofino MPA (Ligurian Sea) with the highest abundance of microplastics (9.67 items/m³). High concentrations of phthalates (1.00–4.32 ng/g fw) were detected in the neustonic/planktonic samples. Phthalates were in bubbler of stranded fin whales, suggesting that they can be used as a tracer of the intake of microplastics.	[37]
Mediterranean Sea		Evaluation of phthalate levels in this species	Presence of phthalates in bubbler (1.48–377.82 ng/g lipid basis). This species can be a potential bioindicator of the presence of microplastics in pelagic environments.	[38]
Ireland	*Mesoplodon mirus* (beaked whale)	Evaluation of exposures trough the analysis of stomach and gut contents	Presence of microplastics in stomachs. Top oceanic predatory species are exposed to plastics; exposure pathways still unclear.	[39]
Mediterranean Sea	*Cetorhinus maximus* (basking shark)	Evaluation of the exposure to phthalates	High concentrations of phthalates in muscle (11.17–156.67 ng/g lipid basis). This species can be a potential bioindicator of microplastics in pelagic environments.	[38]

(*Continued*)

TABLE 4.1 (*Continued*)   Collection of Some Microplastics Exposure and Effects in Animal Species under Both Natural and Laboratorial Conditions

Local	Species	Goals	Main Results	References
Southwest of Plymouth, United Kingdom	*Merlangius merlangus* (whiting fish)	Assessment of plastic ingestion (The study documents microplastics in 10 species of fish from the English Channel.)	Microplastics ingestion (<30 pieces/particles). Presence of polyamide, semi-synthetic cellulosic material and rayon in gastrointestinal tracts).	[36]
Southwest of Plymouth, United Kingdom	*Micromesistius poutassou* (blue whiting fish)	Assessment of plastic ingestion (The study documents microplastics in 10 species of fish from the English Channel.)	Microplastics ingestion (~30 pieces). Presence of polyamide, semi-synthetic cellulosic material and rayon in gastrointestinal tracts).	[36]
Southwest of Plymouth, United Kingdom	*Trisopterus minutus* (poor cod fish)	Assessment of plastic ingestion (The study documents microplastics in 10 species of fish from the English Channel.)	Microplastics ingestion (~40 pieces/particles). Presence of polyamide, semi-synthetic cellulosic material and rayon in gastrointestinal tracts).	[36]
Central Mediterranean Sea	*Xiphias gladius* (swordfish), *Thunnus alalunga* (tuna albacore) and *Thunnus thynnus* (tuna fish)	Evaluation of the presence of plastic debris in stomach	Microplastics ingestion: 29 particles were found in the stomachs of 22 fish. Plastic fragments with different colors and shapes. Swordfish: dominance of mesoplastics (44.4%); albacore: dominance of microplastics (75%); tuna fish: meso- and macroplastics ingested in the same proportion. A relation between fish size and plastic size was found.	[40]
Southwest of Plymouth, United Kingdom	*Zeus faber* (fish)	Assessment of plastics ingestion (The study documents microplastics in 10 species of fish from the English Channel.)	Microplastics ingestion (<60 pieces/particles). Presence of polyamide, semi-synthetic cellulosic material and rayon in gastrointestinal tracts).	[36]

REFERENCES

1. Baptista Neto JA, Wallner-Kersanach M, and Patchineelam SM. Poluição Marinha. In: *Marinha. Editora Interferência.* Rio de Janeiro, Brazil, 2008.
2. Clark BR. *Marine Pollution.* Claredon Press, United Kingdom, 1997.
3. UNEP. Marine Litter, An Analytical Overview. 2005.
4. Kennish JM. *Practical Handbook of Estuarine and Marine Pollution.* CRC Press, Boca Raton, 1997, p. 524.
5. Carpenter EJ, Anderson SJ, Harvey GR, Miklas HP, and Peck BB. Polystyrene spherules in coastal waters. *Science.* 1972, 178, 749–750.
6. Barnes DKA, Galgani F, Thompson RC, and Barlaz M. Environmental accumulation and fragmentation of plastic debris in global. *Phil TR Soc B* 2009, 364, 1985–1998.
7. Thompson RC, Swan SH, Moore CJ, and vom Saal FS. Our plastic age. *Phil TR Soc B* 2009, 364, 1973–1976.
8. Andrady AL. Microplastics in the marine environment. *Mar Pollut Bull.* 2011, 62, 1596–1605.
9. Hinojosa I, and Thiel M. Floating marine debris in fjords, gulfs and channels of southern Chile. *Mar Pollut Bull.*, 2009, 58, 341–350.
10. Cole M, Lindeque P, Halsband C, and Galloway TS. Microplastics as contaminants in the marine environment: A review. *Mar Pollut Bull.* 2011, 62, 2588–2597.
11. Bakir A, Rowland SJ, and Thompson RC. Competitive sorption of persistent organic pollutants onto microplastics in the marine environment. *Mar Pollut Bull.* 2012, 64, 2782–2789.
12. Wright SL, Thompson RC, and Gallowaya TS. The physical impacts of microplastics on marine organisms: A review. *Environ Pollut.* 2013, 178, 483–492.
13. Ivar do Sul JA, and Costa MF. The present and future of microplastic pollution in the marine environment. *Environ Pollution* 2014, 185, 352–364.
14. Cauwenberghe LV, Devriese L, Galgani F, Robbens J, and Janssen CR. Microplastics in sediments: A review of techniques, occurrence and effects. *Mar Environ Res.* 2015,111:5–17.
15. Derraik JGB. The pollution of the marine environment by plastic debris: A review. *Mar Pollut Bull.* 2002, 44, 842–852.
16. Browne MA, Dissanayake A, Galloway TS, Lowe DM, and Thompson RC. Ingested microscopic plastic translocates to the circulatory system of the mussel Mytilus edulis (L.). *Environ Sci Technol.* 2008, 42, 5026–5031.
17. Carson HS, Colbert SL, Kaylor MJ, and McDermid KJ. Small plastic debris changes water movement and heat transfer through beach sediments. *Mar Pollut Bull.* 2011, 62, 1708–1713.
18. Yntema CL, and Mrosovsky N. Critical periods and pivotal temperatures for sexual differentiation in loggerhead sea turtles. *Canadian J Zool* 1982, 60 1012–1016.
19. Penn D, and Brockmann HJ. Nest-site selection in the horseshoe-crab, Limulus polyphemus. *Biol Bull.* 1994, 187, 373–384.
20. D'avila S, and Bessa ECD. Influence of moisture on growth and egg production by *Subulina octona* (Bruguiere) (Mollusca, Subulinidae), reared in different substrates, under laboratorial conditions. *Rev Bras Zool.* 2005, 22, 349–353.
21. Di Domenico M, Lana PD, and Garraffoni ARS. Distribution patterns of interstitial polychaetes in sandy beaches of southern Brazil. *Marine Ecol an Evol Perspect* 2009, 30, 47–62.

22. Quinn T. Habitat characteristics of an intertidal aggregation of Pacific sandlance (Ammodytes hexapterus) at a North Puget Sound Beach in Washington. *Northwest Sci.* 1999, 73, 44–49.

23. Evans SM. Tributyltin pollution: The catastrophe that never happened. *Mar Pollut Bullet* 1999, 38, 629–636.

24. Collignon A, Hecq JH, Glagani F, Voisin P, Collard F, Goffart A. Neustonic microplastic and zooplankton in the North Western Mediterranean Sea. *Mar Pollut Bullet* 2012, 64, 861–864.

25. Nobre CR, Santana MFM, Maluf A, Cortez FS, Cesar A, Pereira CDS, and Turra A. Assessment of microplastic toxicity to embryonic development of the sea urchin Lytechinus variegatus (Echinodermata: Echinoidea). *Mar Pollut Bullet* 2015, 92, 99–104.

26. van Cauwenberghe L, Claessens M, Vandegehuchte MB, and Janssen CR. Microplastics are taken up by mussels (Mytilus edulis) and lugworms (Arenicola marina) living in natural habitats. *Environ Pollut.* 2015, 199, 10–17.

27. Moos NV, Burkhardt-Holm P, and Köhler A. Uptake and effects of microplastics on cells and tissue of the blue mussel Mytilus edulis L. after an experimental exposure. *Environ Sci and Technol* 2012, 46, 11327–11335.

28. Wegner A, Besseling E, Foekema EM, Kamermans P, and Koelmans AA. Effects of nanopolystyrene on the feeding behavior of the blue mussel (Mytilus edulis L.). *Environ Toxicol and Chem* 2012, 31, 2490–2497.

29. Browne M, Dissanayake AD, Galloway TS, Lowe DM, and Thompson RC. Ingested microscopic plastic translocates to the circulatory system of the mussel, Mytilus edulis (L.). *Environ Sci and Technol* 2008, 42, 5026–5031.

30. van Cauwenberghe L, and Janssen CR. Microplastics in bivalves cultured for human consumption. *Environ Pollut.* 2014, 193, 65–70.

31. Avio CG, Gorbi E, Milan M, Benedetti M, Fattorini D, d'Errico G, Pauletto M et al. Pollutants bioavailability and toxicological risk from microplastics to marine mussels. *Environ Pollut.* 2015, 198, 211–222.

32. Besseling E, Wegner A, Foekema EM, Martine J, van den Heuvel-Greve, and Koelmans AA. Effects of microplastic on fitness and PCB bioaccumulation by the lugworm Arenicola marina (L.). *Environ Sci Technol.* 2013, 2, 593–600.

33. Wright SL, Rowe D, Thompson RC, and Galloway TS. Microplastic ingestion decreases energy reserves in marine worms. *Curr Biol.* 2013, 23, 1031–1033.

34. Lusher AL, McHugh M, and Thompson RC. 2013. Occurrence of microplastics in the gastrointestinal tract of pelagic and demersal fish from the English Channel. *Mar Pollut Bullet* 2013, 67, 94–99.

35. Carlos de Sá L, Luís LG, and Guilhermino L. Effects of microplastics on juveniles of the common goby (Pomatoschistus microps): Confusion with prey, reduction of the predatory performance and efficiency, and possible influence of developmental conditions. *Environ Pollut.* 2015, 196, 359–362.

36. Luís LG, Ferreira P, Fonte E, Oliveira M, and Guilhermino L. Does the presence of microplastics influence the acute toxicity of chromium (VI) to early juveniles of the common goby (Pomatoschistus microps)? A study with juveniles from two wild estuarine populations. *Aquat Toxicol* 2015, 164, 163–174.

37. Fossi MC, Panti C, Guerranti C, Coppola D, Giannetti M, Marsili L, Minutoli R. Are baleen whales exposed to the threat of microplastics? A case study of the Mediterranean fin whale (Balaenoptera physalus). *Mar Pollut Bullet* 2012, 64, 2374–2379.

38. Fossi MC, Coppola D, Baini M, Giannetti M, Guerranti C, Marsili L, Panti C et al. Large filter feeding marine organisms as indicators of microplastic in the pelagic environment: The case studies of the mediterranean basking shark (Cetorhinus maximus) and fin whale (Balaenoptera physalus). *Mar Environ Res.* 2014, 100, 17–24.

39. Lusher AL, Hernandez-Milian G, O'Brien J, Berrow S, O'Connor I, Officer R. Microplastic and macroplastic ingestion by a deep diving, oceanic cetacean: The true's beaked whale Mesoplodon mirus. *Environ Pollut.* 2015, 199, 185–191.

40. Romeo T, Pietro B, Peda C, Consoli P, Andaloro F, Fossi MC. First evidence of presence of plastic debris in stomach of large pelagic fish in the Mediterranean Sea. *Mar Pollut Bullet* 2015, 95, 358–361.

CHAPTER 5

Impacts of Macro- and Microplastics on Macrozoobenthos Abundance in the Intertidal Zone*

A. P. Bangun, H. Wahyuningsih and A. Muhtadi

CONTENTS

5.1 INTRODUCTION

Development is an inseparable part of human civilization. Life without development activity is a setback in human civilization. Often, developments such as industrial and agricultural factories do not pay attention to the environmental aspect, thus destroying nature. The plastics industry is one of the industries that continues to grow along with the high use of plastics in society [1,2]. The low public awareness of plastic waste management becomes one of the causes of the amount of plastic waste in the environment, especially in coastal waters.

Marine debris is a persistent solid, manufactured or processed by humans, directly or indirectly, intentionally or unintentionally, disposed of or abandoned in the marine environment. Type of marine waste including plastic, cloth, foam, Styrofoam, glass, ceramics, metal, paper, rubber and wood [3,4]. The size category is used to classify marine

* Reprinted from A P Bangun et al. 2018. *IOP Conf. Ser.: Earth Environ. Sci.* 122, 012102. Content from this work may be used under the terms of the Creative Commons Attribution 3.0 licence. Any further distribution of this work must maintain attribution to the author(s) and the title of the work, journal citation and DOI. Published under licence by IOP Publishing Ltd. This is an open access article.

debris, i.e., megadebris (>100 mm), macrodebris (>20–100 mm), mesodebris (>5–20 mm), and microdebris (0.3–5 mm) [1,3,4].

Various problems arise due to marine debris, such as reduced coastal beauty, origin of various diseases, effects on the food web and reduced productivity of fish caught [3,5–8]. The potential effects of marine waste chemically tend to increase as the size of plastic particles (microplastics) decreases, while the physical effect increases with the increase in macro-debris size [3]. Macrodebris provides a physical impact, such as closing the surface of the sediment and disrupting the movement of aquatic organisms [6,9,10], and can prevent the growth of mangrove seeds [9]. Macroplastics can also be a host for the growth of invasive species, especially those attached to the substrate [6,8].

Microplastics potentially threaten more seriously than large plastic materials, as organisms that inhabit lower tropical levels, such as planktons that have particles susceptible to microplastic digestion processes, as a result, can affect high-level tropical organisms through bioaccumulation [6,10,11]. Microplastics are consumed by marine organisms when one of the microplastic particles resembles food [8,12–14].

The presence of macro- and microplastics in the environment, especially the aquatic environment, can have an impact on the existence of organisms [7,8]. Macrozoobenthos is one of the several organisms that receive pressure from plastic waste contamination. Macrozoobenthos is often used to predict the imbalance of the physical, chemical and biological environments of waters [15]. The polluted waters will affect the viability of macrozoobenthos organisms because macrozoobenthos is a water biota that is easily affected by the presence of pollutants.

The phenomenon of marine waste in the form of plastic causes unrest in the community with the presence of garbage that has polluted coastal and marine areas, including the Village Jaring Halus Langkat. In addition, the absence of preliminary information about microplastics in this region is one of the constraints to manage the potential of fisheries and marine-based environmentally friendliness. Based on this, it is necessary to conduct a study to find out the macroplastic and microplastic distribution and its impact on macrozoobenthos in the Jaring Halus Village.

5.2 MATERIALS AND METHODS

5.2.1 Study Site

The research was conducted in Jaring Halus Village, Langkat Regency, North Sumatra Province, Indonesia. Data were collected from February to April 2017. The tools used were global position system (GPS), meter, sample bottle, tool box, core, 1 m × 1 m board, filter, shovel, digital camera and stationery.

5.2.2 Method of Macroplastic Survey

The macrodebric samples (>20 mm) were collected with transects (1 × 1 m) from each substation with the highest tidal boundaries and the lowest tidal boundaries divided into three sections [9]. The macrodebric composition were grouped into plastics, fabrics, foams, Styrofoam, glass, metal, rubber and wood. Samples were collected into sacks and labeled. The items (to further explain the flakes) in each macrodebric group were collected macroplastic, dried, calculated and weighed. The parameters taken include the number of items (item m^{-2}) and weight (g m^{-2}) [16].

5.2.3 Method of Microplastic Survey

Sediment sampling (1 L) was performed with the core based on three stratified depths (0–30 cm). The core placement is performed on three sections (top, middle and bottom edge) at the substations at the highest tide and lows. The separation of microplastic particles (0.045–5 mm) from the sediments was carried out by several stages, namely (a) drying, (b) volume reduction, (c) separation of density, (d) filtration and (e) visual sorting. Drying was done with a 105°C oven for 72 hours. The dry-sediment volume reduction step was performed by filtration (5 mm in size) [17]. The density separation step was carried out by mixing the dry sediment sample (1 kg) and the saturated NaCl (3 L) solution, and the mixture are stirred for 2 minutes [18].

Floating plastics were polystyrene, polyethylene and polypropylene. The filtration stage was carried out by filtering the supernatant (size 45 μm). Microplastic particles were visualized using a monocular microscope and grouped into four types, namely film, fiber, fragments and pellets. The parameters taken were density (particle/kg dry sediment) [17].

5.2.4 Sampling of Macrozoobenthos

A sampling of macrozoobenthos was done on the same plot/transect as macro- and microplastic retrieval. Macrozoobenthos were taken on each transect plot with a depth of 30 cm. All the substrate on the plot was removed with a shovel, then stored in a plastic bag for separation. The separation between macrozoobenthos and substrate was done in the field with the help of water and filter. Samples of macrozoobenthos that had been separated from the substrate were fed into a 70% alcohol-treated sample bottle to be identified in the laboratory.

5.2.5 Data Analysis

The macroplastic abundance is calculated by the formula [4]:

$$K = \frac{n}{w \times l} \tag{5.1}$$

Information:

K = Macroplastic Abundance
n = Number of Macroplastic (item/m^3)
w = Width of Sampling Area
l = Length of Sampling Area

Microplastic abundance is calculated by the formula [4]:

$$K = \frac{n}{a \times b} \tag{5.2}$$

Information:

K = Microplastic Abundance
N = Number of Microplastic (item/m^3)
A = Sample Area of Sampling
H = Depth of Sampling

Biota density of macrozoobenthos with formula [15]:

$$K = \frac{ni}{A}$$

(5.3)

Information:

K = Density
ni = Number of Individuals of a Type
A = Area

Correlation analysis between macro- and microplastic to macrozoobenthos was done using Pearson correlation SPSS.

5.3 RESULTS AND DISCUSSIONS

5.3.1 Macroplastics

The total number of macroplastics collected from three stations divided into 27 observation plots were 308 items with a total weight is 3689.87 g. The highest macroplastic density found in station 2 ranged between 18.33 and 190.33 species/m^2 with weights ranging from 246.33 to 2103 g. The lowest density found in station 3 ranged from 3.33 to 11.67 species/m^2 with weights ranging from 13.46 to 117.67 g. Density and macroplastic weights can be seen in Table 5.1.

Plastics is one of the most common types of waste found in various places both on land and in waters. The common marine waste category found in the Coastal Village of Jaring Halus is the macroplastic, i.e., plastic waste with a size larger than 5 mm, with as many as 308 items (Table 5.1). Hastuti (2014) found macroplastic density in Mangrove Ecosystem Pantai Indah Kapuk Jakarta as many as 6079 species. Some research results abroad have found that marine debris is dominated by macroplastic at 48% in Cassina

TABLE 5.1   Abundance and Macroplastic Weights

Station	Macroplastic Abundance (Item/m²)				Macroplastic Weight (g)			
	Plot 1	Plot 2	Plot 3	Total	Plot 1	Plot 2	Plot 3	Total
1	32.67	9.67	5	47	393	115.67	65.67	574
2	190.33	31.67	18.33	240	2203	508	246.33	2957
3	11.67	5.33	3.33	20	113.67	31.08	13.46	158
Total	234.67	46.67	26.67	308	2709.67	654.75	325.46	3689.87

Beach, Brazil [19]; 67.6% in Northeast Coast Brazil [20]; 89% in Bootless Bay, Papua, New Guinea [21]; 91% in Midway, North Pacific [22]; 68% in Monterey Beach, US [23]; 77% in Kaohsiung, Taiwan [24]; 45.2%–95% in Eastern Mediterranean and Black Seas [25] and 60 5%–80% in Gulf Coast of Guinea of Ghana [26]. The macroplastic proportion is dominant because its density is lower than glass, metal and water density so it is easily transportable [1,7,25].

The highest macroplastic densities found at station 2 (near community settlements) ranged from 18.33 to 190.33 species/m^2. This is allegedly due to community activity that contributes most macroplastic compared to station 1 (faced with Malacca Strait). Macroplastic density is influenced by the amount of marine debris carried from the sea, such as fishing and shipping activities in the Malacca Strait and station 3 (opposite the Ular River) where the macroplastic density is influenced by the amount of waste carried by the river. It also indicates that the station distance to the main pollutant source (settlement) effects macroplastic density.

The result of the Kruskal–Wallis test shows that macroplastic density between stations is significantly different with a p-value = 0.005. Mann–Whitney test results show that the macroplastic density of station 1 is significantly different from station 2 with a p-value = 0.031 and not significantly different from station 3 with a p-value = 0.147. The macroplastic density at station 2 differs significantly from station 3 with p = 0.02. The result of the Kruskal–Wallis test shows that macroplastic density between plots is significantly different with p = 0.004. The Mann–Whitney test results showed that the macroplastic density of plot 1 was significantly different from plot 2 with p = 0.031 and significantly different from plot 3 with p = 0.003. The macroplastic density of plot 2 is not significantly different from plot 3 with p = 0.63.

The result of the Kruskal–Wallis test shows that macroplastic density between station plot of observation is not significantly different with p = 0.857. This indicates that the highest tidal distance and lows at each station do not affect macroplastic density. This is in accordance with the results of other studies [27], which obtain macrodebris density in the mangrove ecosystem of Pantai Indah Kapuk is not significantly different between stations and indicates that the station distance from the sea does not affect macrodebris density.

5.3.2 Microplastic

The highest microplastic density is found in plot 1 of all observation stations, which is the highest tidal boundary. Density decreases in the lowest tide and lows. This indicates that in the intertidal zone, the microplastic decreases in density as the distance to the sea increases. Proven similarly is that the microplastic density in tidal zones at the highest tidal boundary is higher than at the lowest tidal boundary, and there is a real difference between the two [14,28]. The zone at the lowest tide is a very dynamic zone; deposition can occur constantly. Sediments in the upper layers in this zone are susceptible to runoff and become suspended again.

Microplastic film type is the highest type found in all observation plots. The film comes from fragmentation of plastic waste such as packets of food and soft drinks, plastic bags and plastic wrapping commonly used in fishing activities. The film also has the lowest density, so it is easily distributed by the presence of currents and tides [18,28]. Other research results on the Belgian Coast [28] gained microplastic fiber type flavors (59%). While in Muara Badak, East Kalimantan [29], the highest abundance of microplastic fragments (58.97%) was found. This indicates the type of microplastic depends on the

TABLE 5.2   The Percentage of Density and Average Density of Microplastic

| Station | Film | Percentage of Microplastic Abundance (%) | | | Abundance Average (Item/kg) |
		Fiber	Fragment	Pellet	
1	60.27	22.30	17.47	0	89
2	45.07	30.03	24.63	0.3	173.33
3	51.57	22.30	26.13	0	64
Total	52.30	24.88	22.74	0.1	326.33

macroplastic and the source of the plastic pollutant. The high fiber on the Belgian coast is due to the point of location close to the fishery port area. Fiber is derived from the activity of catching while the fragment is the result of a piece of plastic product with a very strong synthesis polymer (Table 5.2) [28].

The result of the Kruskal–Wallis test shows that the density of microplastic between stations is significantly different with the value of p = 0.0. This indicates that there is one station with a microplastic density that is significantly different from other stations. Mann–Whitney test results show that the stationary microplastic density station 1 is significantly different from station 2 with p value = 0.031 and not significantly different from station 3 with p = 0.094. The station's microplastic density station 2 differs significantly from station 3 with p = 0.002. This proves that the microplastic density will be higher with reduced distance from the main pollutant source, and with increasing distance from the main pollutant source, the density is more affected by other pollutant sources such as river and sea currents [30]. The result of the Kruskal–Wallis test showed that the microplastic density between the observed plots was not significantly different with the value of p = 0.984.

5.3.3 Macrozoobenthos

The highest density of macrozoobenthos was found in station 3 with a density of 61 ind m^{-2}. In plot 1, station 2, no macrozoobenthos is found in any depth stratification. The average density of macrozoobenthos at the observation point can be seen in Table 5.3.

The result of analysis using Pearson correlation shows the relationship between macrozoobenthos density with macroplastic equal to 0.633 with the level of strong and

TABLE 5.3   The Average Density of Macrozoobenthos

| Stations | Macrozoobenthos Density | | | | | | | | | Total |
| | 1[a] | | | 2[a] | | | 3[a] | | | |
	10[b]	20[b]	30[b]	10[b]	20[b]	30[b]	10[b]	20[b]	30[b]	
1	6	4	4	5	3	3	4	3	4	36
2	0	0	0	3	2	3	3	5	3	18
3	2	13	3	19	5	2	9	4	3	61
Total	8	17	7	26	10	8	16	12	10	115

[a] Plot.
[b] Depth.

microplastic relationship of 0.386 with the low level of relationship. Macroplastic density with microplastic of 0.756 with strong relationship level. This suggests that macroplastic density affects macrozoobenthos density. Increased macroplastic density causes decreased macrozoobenthos density. The macroplastic can close the surface of the sediment and disrupt the movement of aquatic organisms [6,9].

Microplastic density has a low correlation to macrozoobenthos, presumably because microplastic affects the digestive system of macrozoobenthos which can lead to accumulation in macrobenthos digestive organs [7,8,18,30]. Macroplastic density with microplastic of 0.756 with strong relationship level. This indicates that the micro plastic density is affected by the macroplastic density. The higher the macroplastic density will lead to increased microplastic density. Microplasticity results from macroplastic fragmentation [1,7,28,29,31].

5.4 CONCLUSIONS

An increase in the number of macroplastic densities has led to a decrease in macrozoobenthos density. Macroplastic density has a strong negative impact on macrozoobenthos density, while microplastic density has a low negative impact on macrozoobenthos density. Increased macroplastic density leads to an increase in microplastic density.

REFERENCES

1. Hammer J, Parsons J, Kraak MHS. Plastics in the Marine Environment: The Dark Side of a Modern Gift. In: Whitacre DM (ed.) *Reviews of Environmental Contamination and Toxicology.* Springer Science + Business Media, 2012.
2. Thevenon F et al. Plastic Debris in the Ocean: The Characterization of Marine Plastics and their Environmental Impacts, Situation Analysis Report. Gland, IUCN, Switzerland, 2014, 52.
3. United Nations Environment Programme. UNEP Year Book Emerging Issues in Our Global Environment. UNEP, Nairobi, 2011.
4. National Oceanic and Atmospheric Administration. Programmatic Environmental Assessment (PEA) for the NOAA Marine Debris Program (MDP). NOAA, Maryland, 2013.
5. Citrasari N et al. *J Biol Res*, 2012, 18(1), 83–85.
6. Irwin K. The Impacts of Marine Debris: A Review and Synthesis of Existing Research. *Living Oceans Soc* December 2012.
7. Galgani F et al. Global Distribution, Composition and Abundance of Marine Litter. In: Bergmann M et al. (eds.) *Marine Anthropogenic Litter.* 2015, DOI 10.1007/978-3-319-16510-3_2
8. Ogunola OS, and Palanisami T. *J Pollut Eff Cont* 2016, 4, 161.
9. Smith SDA, and Markic A. *PLOS ONE* 2013, 8(12).
10. *National Oceanic and Atmospheric Administration Marine Debris Program 2016 Report on Marine* Debris Impacts on Coastal and Benthic Habitats. NOAA, Silver Spring, MD.
11. Eriksen M et al. Plastic. *PLOS ONE* 2014, 9(12).
12. Browne MA et al. *Environ Sci Technol* 2008, 42(1), 5026–5031.
13. Boerger CM et al. *Mar Pollut Bull* 2010, 60(12), 2275–2278.

14. Cauwenberghe LV et al. *Mar Pollut Bull* 2013, 73(1), 161–169.
15. Odum EP. *Dasar-Dasar Ekologi. Edisi Ketiga.* Gajah Mada University Press, Yogayakarta, 1995.
16. Peters K, and Flaherty T. *Marine Debris in Gulf Saint Vincent Bioregion.* Government of South Australia, Adelaide (AU), Australia, 2011.
17. Hidalgo-Ruz V et al. *Environ Sci Technol* 2012, 46(1), 3060–3075.
18. Claessens M et al. *Mar Pollut Bull* 2011, 62(1), 2199–2204.
19. Tourinho PS, and Fillmann G. *J Integr Coast Manag* 2011, 11, 97–102.
20. Costa MF et al. *J Coast Res* 2011, 64(1), 339–343.
21. Smith SDA. *Mar Pollut Bull* 2012, 64(9), 1880–1883.
22. Ribic C et al. *Mar Pollut Bull* 2012, 64(8), 1726–1729.
23. Rosevelt C et al. *Mar Pollut Bull* 2013, 71(1–2), 299–306.
24. Liu T et al. *Mar Pollut Bull* 2013, 72, 99–106.
25. Ioakeimidis C et al. *Mar Pollut Bull* 2014, 89(1–2), 296–304.
26. Van Dyck IP et al. *J Geosci Environ Prot* 2016, 4, 21–36.
27. Hastuti AR. Distribusi Spasial Sampah Laut di Ekosistem Mangrove Pantai Indah Kapuk Jakarta. Skripsi. FPIK. Institut Pertanian Bogor. Bogor, 2014.
28. Syakti AD, Bouhroum R, Hidayati NV, Koenawan CJ, Boulkamh A, Sulistyo I. Beach macro-litter monitoring and floating microplastic in a coastal area of Indonesia. *Mar Pollut Bull* 2017, 122(1–2), 217–225.
29. Dewi IS. *Depik* 2015, 4(3), 121–131.
30. Andrady AL. *Mar Pollut Bull* 2011, 62, 1596–1605.
31. Cordova MR, and Wahyudi AJ. *Mar Res Indonesia* 2016, 41(1), 27–35.

Plastics in Food

CHAPTER 6

Microplastics and Nanoplastics in Food

Mohammed Asadullah Jahangir, Sadaf Jamal Gilani,
Abdul Muheem and Syed Sarim Imam

CONTENTS

6.1 INTRODUCTION

Plastics were first produced in the early 20th century. The most common form of plastics, that is, polyethylene soon became a boon for mankind, though this assumption did not last for long. As soon as it was realized that plastics do not decompose naturally, the environmental issues and its health hazards started rising. Because of their enormous popularity, plastics can now be found everywhere. Being an integral part of almost everyone's life, plastics became the worst manmade environmental issue of all time. Being present in almost every household, plastics helped humans in preserving foods, etc. However, its overuse caused millions of tons of plastic wastes every year [1].

Pollution by plastic component is quite evident. It can be easily seen at landfills and seashores. Although plastics wastes cause environmental issues, the real threat comes from the minute fragments which are produced from the breakdown of large fragments of plastics. These small fragments are usually classified into micro- and nanoplastics. Being in the size range of a few micrometers, they are hard to be recognized by the naked eyes. It is very difficult to isolate them from their micro-environment by simple methods [1].

As per an estimate by the UN in the year 2017, there are about 51,000 billion particles of plastics in the sea, a number which is almost 500 times more than the total number of stars present in our Milky Way galaxy [2]. Plastic in its fragmented form is adversely affecting the marine ecosystem. A global interest has arisen to tackle the hazards of plastic wastes in seas and waterways which are adversely affecting wildlife and natural habitats. The floating plastics or plastic soups cause debris to be formed which are fragmented into micro- and nanoplastics.

6.2 CLASSIFICATION OF FRAGMENTED PLASTICS

Fragmented plastics are usually classified into microplastics and nanoplastics depending on their size.

6.2.1 Microplastics

Microplastics are fragments of synthetic polymers having the upper limit of 5 mm. There is no official lower limit of size. They usually have variable shapes but mostly present as fibers. Their chemical composition, color, density and other characteristics are also variable [3,4]. Microplastics are further classified into primary and secondary forms.

Primary microplastics are usually manufactured in industries as scrubbers and are used to blast clean surfaces. They are also manufactured in the form of plastic powders and used in molding and as microbeads in cosmetic formulations [5,6]. Secondary microplastics are the predominant form which are formed by the fragmentation of plastic debris in the oceans. This fragmentation process occurs owing to the exposure of plastics to ultraviolet radiation and by physical abrasion. They can originate from both sea and land-based sources. Land sources of microplastics include packaging materials, polyethylene bags and industrial wastes. Sea sources are mainly the sewage released from humans, industries, ships, fishing materials and equipment. Biofouling of these plastic fragments make them to sink to the depth of the sea floor. Microplastics from terrestrial sources consist of wastes including personal-care products such as cleaning agents, toothpastes, textile fibers. They get transported to the sewage system. The current sewage system is not as potent to clean

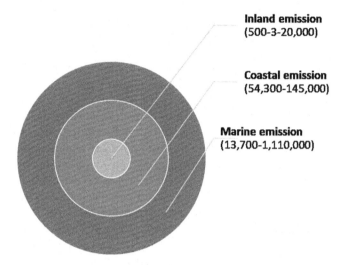

Inland emission
(500-3-20,000)

Coastal emission
(54,300-145,000)

Marine emission
(13,700-1,110,000)

FIGURE 6.1   Secondary microplastic emission by different sources.

these fragmented plastic forms, and thus, they eventually get deposited into the marine ecosystem [6,7].

Zooplankton, invertebrates and vertebrates are exposed to microplastics through biomagnification. As per an estimate [7], the amount of secondary microplastic emission to the marine ecosystem falls between 68,5000 and 275,000 tonnes per year. These data can be further classified into coastal, inland and marine emission (Figure 6.1). Among all polymers, the most commonly found are polyethylene, polystyrene and polypropylene. Additives of both organic and inorganic nature such as phthalates, bisphenol A and alkylphenols and inorganic additives such as barium, sulfur, titanium dioxide and zinc make up to 4% of the weight of plastics [8,9].

Microplastics can travel long distances from their place of origin through air currents and get deposited on land and water. Although their impact on human and marine life is obvious, there is no ecological or toxicological threat identified so far from the chemical and physical composition of microplastics. Although indirect threats are quite evident, microplastics are usually hydrophobic in nature [3] and thus adsorb harmful organic and organochlorine pesticides such as polychlorinated biphenyls (PCBs), polycyclic aromatic hydrocarbons (PAHs) and dichlorodiphenyltrichloroethane (DDT). In the food chain, these microplastics are bioaccumulated.

6.2.2 Nanoplastics

As per EFSA, nanoplastics measure from 0.001 to 0.1 μm (i.e., 1–100 nm). Different theories are present about the formation of nanoplastics. It is assumed that fragmentation of plastic debris over the years may lead to formation of nanoparticles [10,11]. In an experiment conducted by Lambert and Wagner, nanoplastics were formed by degradation of polystyrene disposable coffee cup lids [12]. Microbial degradation is also expected to play its role in hydrocarbon degrading, as several such microorganisms have been identified. They used to thrive in the plastic debris environment in the oceans [13]. Continuous fragmentation of microplastics into nanoplastics also occurs in nature [14]. Industrial nanoplastics also accumulate in oceans as effluents.

6.3 IMPACT OF MICRO- AND NANOPLASTICS
ON AQUACULTURE AND ITS PRODUCTS

Aquatic animals of commercial or non-commercial use unintentionally ingest micro- and nanoplastics, which are dumped into oceans and inland waters. These plastic fragments along with the adsorbed chemicals pose a threat to food safety of aquaculture products and fishery itself [15]. Plastics contaminate aquatic environments either by direct runoff or by degradation of macro- or meso-plastic debris. Direct release results from industrial wastes, untreated sewage and fishing gears. Out of 25 most important aquatic species, microplastics have been found in 12 genera and, thus, have a global impact on marine fisheries [15]. They therefore pose an extreme food-safety concern, as some of the micro- and nanoplastic components along with the additives and adsorbed polymers are carcinogenic in nature. These micro- and nanoplastics are capable of causing negative effects like alteration in lipid metabolism, physical damage, cytotoxicity and change in behavior in fishes. Ingestion of micro- and nanoplastics blocks the digestive system [16], which may cause either starvation or satiation in fishes. It also slows down the digestive process in aquatic animals. As per the study conducted by Mazurais et al., under laboratory conditions, the mortality rate of young fishes feeding on microplastics was found to be significantly higher when compared to the control [17]. Mechanical injuries and ulceration are also caused by microplastics with sharp edges. In another study by Pedia et al., fishes fed with microplastics under laboratory conditions showed alteration in histopathology of the distal intestine like detachment of mucosal epithelium form lamina propria, swelling and shortening of villi, increase in the number of goblet cells, vacuolation of enterocytes, loss of regular structure of serosa, widened lamina propria and hyperplasia of goblet cells [18].

Micro- and nanoplastics often enter the fish circulatory system. Nanoplastics are capable of changing fish metabolism through alteration of the triglycerides-to-cholesterol ratio in blood serum and through distribution of cholesterol between liver and muscle [19]. Downregulation of amino acids and upregulation of fatty acids was also observed upon exposure of fishes to micro- and nanoparticles [20]. They also reported necrosis, accumulation of lipid droplets and infiltration upon exposure of zebrafish liver to microplastics. In another study by Rochman et al., it was concluded that exposure of fish liver to microplastics shows signs of fatty vacuolation, necrosis, stress and glycogen depletion [21].

6.4 SOURCES OF HUMAN CONSUMPTION
OF FRAGMENTED PLASTICS

There are various sources through which humans are consuming micro- and nanoplastics. They can be broadly classified into marine and non-marine source.

6.4.1 Marine Source of Consumption

Biomagnification and bioaccumulation are the major phenomena through which micro and nanoplastics reach higher levels of the food chain [90]. However, consumption of fishes and shellfishes are not the only source of fragmented plastics. It is quite difficult to conclude the actual amount of microplastics being consumed from fishes because of the limited data available. Most of the studies are limited to analysis of stomach or gut content [22]. Microplastics have also been reported in canned fish. Apart from that, sea salt is another source of microplastic. One kilogram of sea salt is expected to contain over 600 microplastics, but this number increases depending upon the consumption of salt.

6.4.2 Non-Marine Source of Consumption

Globally, the intake of microplastics is far more from non-marine sources than the sea. Land animals also eat a lot of microplastics from different sources. Scientists have also confirmed the presence of microplastics in chickens, honey and beer, nevertheless the biggest source of microplastics consumption is bottled water. A study between single-use and reused water bottles showed that reused water bottles contain 5–14-fold more microplastics than single-use water bottles. There is evidence of microparticles being present in huge amounts in indoor dust [22].

6.4.3 Inhalation Exposure to Fragmented Plastics

Debris and municipal effluents usually contain fragmented plastics. They are retained in sludge and reach agricultural lands in the form of fertilizer. The dried sludge-based fertilizers are transported through wind [23]. In a recent study, it was reported that microplastics are also present in the atmospheric dust [24]. Out of the total fallout studied, 30% were plastics and fibers. Densely populated areas were found to contain more microplastics than less-populated areas. Thus, humans living in urban areas are at higher risk of inhalational exposure to microplastics [24]. In another study, it was confirmed that synthetic rubber or tires contain microplastics. Abrasion of such materials has been reported to contain particulate matter. Exposure concentration and size of the fragmented plastics determines the potential risk. Once the micro- and nanoplastics or fibers enter into the respiratory tract, they are likely to be trapped in the lung lining fluid. However, in individuals with impaired mechanism of defense, these fibers may reach the lungs.

6.4.4 Uptake and Translocation of Fragmented Plastics

Individuals working in textile or other industries capable of producing microplastics are supposed to be at an occupational health risk. Common issues reported by such individuals are respiratory irritation [25], dyspnea, coughing and reduced lung capacity [26,27]. Histopathological studies of lung biopsies of workers of textile industries showed granulomatous lesions and interstitial fibrosis due to fragmented textile dust. The symptoms were similar to that of allergic alveolitis [28]. It has been reported that malignant and non-neoplastic lung tissue contains both plastic and cellulose microfibers [29]. Size and shape of the fibers play a key role in toxicity. Small and thin fibers are capable of passing through the respiratory tract and cause toxicity to pulmonary cells [25]. Nylon fibers of less than 2 μm diameter can be respired.

6.4.4.1 Uptake in Respiratory Tract by Endocytosis

Microplastics may enter the respiratory epithelium via diffusion. However, active cellular uptake has also been reported to transport micro- and nanofragments [30]. Energy-dependent endocytosis and phagocytosis also assist in the active uptake of micro- and nanoparticles [31]. Particles of 1–3 μm are cleared from alveoli through phagocytosis [32].

6.4.4.2 Uptake in Gastrointestinal Tract by Endocytosis

The ileum, specifically the Peyer's patches, are the regions of small intestine and the major sites for uptake and translocation of fragmented particles [33,34]. Other non-biodegradable microelements like titanium dioxide and aluminosilicates are retained over the phagocytic cells of ileum [33]. This region acts as a compartment where microplastics accumulate and can take the endogenous route, thus compromising local immunity.

6.4.4.3 Uptake by the Persorption Method

The persorption method is the process where mechanical kneading of microparticles occurs and they are eventually taken up by the gaps in the epithelial cells into the circulatory system [35,36]. It was observed that micro PVC particles of 5–110 μm tend to pass between the enterocytes in the villi [37]. Persorption has also been reported in humans. Granules have been reported in bile, urine, cerebrospinal fluid, breast milk and peritoneal fluid [38].

6.5 FRAGMENTED PLASTICS IN FOOD AND FOOD PACKAGING

Studies have shown the presence of micro- and nanoplastics in fishes and sea mammals. Table 6.1 summarizes the list of research on food and food products. As per the reports published, the number of various species contaminated with fragmented plastics is very high, with more than 690 aquatic species of both edible and non-edible nature [49]. Although the chances of non-edible species reaching the human diet is very remote, their presence in the ecological balance cannot be denied. Only 200 edible species from marine sources and one from a terrestrial source has been considered [89]. Only one study is reported, by Lwanga et al., that concludes chicken gizzards to contain micro- or nanoplastics [50].

It has been shown that plastic fragments are present in 35% of the plankton eating fishes. On average, 2.1 fragments per fish were found [51]. In a study of Brazilian estuaries by Possatto et al., 18%–33% of catfish showed to have fragmented plastic particles in their stomachs [52]. It has been reported that the fragmented plastics can sediment at the bottom of the sea and species living in the dark zone of oceans are also susceptible to consuming micro- and nanoplastics [53].

Hollman et al. reported that in the benthic region of the sea, mostly crustaceans and polychaetes live. He also concluded that a trophic level transfer and bioaccumulation may also work here [54].

It is common to use fish guts for preparing animal feed, mostly in the poultry farms. Although, there is no report about possible contamination of animal husbandry, which uses fish waste products as animal feed. This area has a scope of research and the possibility of gaining some vital data about contaminated edible animals reaching the human diet. According to the European Food Safety Authority (EFSA) [5], blue mussels (*Mytilus edulis*) cultivated for human consumption are exposed to microplastics of 2–10 μm. This is in accordance with the conclusion of Browne et al. [55]. The EFSA report also concludes about the possible presence of microplastics in bivalves, fishes and shrimps. Presence of microplastics in common mussels was also reported by Li et al. [56–58]. In their study, Li and co-workers concluded that the contamination level was higher in wild mussels than farmed ones, and processed mussels were more contaminated with microplastics than nonprocessed mussels [56].

Liebezeit released two research studies reporting contamination of honey by micro- and nanoplastics in samples from producers and supermarkets. The first of the two studies concluded an average presence of 166±147 fibers/kg and 9±9 fragments/kg of honey

TABLE 6.1   List of Reported Studies on Food Contamination by Fragmented Plastics

Food/Food Product	Level of Microplastic Contamination	Inference	References
Honey	Average of 166 ± 147 fibers/kg and 9 ± 9 fragments/kg	Contamination by cellulose fibers, chitin fragments	[39,40]
Honey samples from Germany	40–660 fibers/kg of honey and 0–38 fragments/kg of honey	Contamination by pollen type string-like fibers and chitin	[40]
Honey samples from Switzerland	Black carbon particles (1760–8680 particles/kg); Cellulose based fibers (32–108 fibers/kg); Pollen particles (8–64 particles/kg); Cellulose or chitin fibers (132–728 fibers/kg); glass particles (60–172 particles/kg)	Contamination by carbon particles, cellulose fibers, chitin fibers, glass particles	[41]
Turkish salts	Turkish sea salt: (16–84 particles/kg); Turkish lake salt (8–102 particles/kg); Turkish rock salt (9–16 particles/kg)	Contamination by polyethylene fragments (22.9%) and polypropylene fragments (19.2%)	[42]
Commercialized salts available in US stores	46.7–806 particles/kg		[43]
Salt brands from Malaysian market	1–10 particles/kg	Contamination by plastic polymer fragments	[44]
Sea, lake and rock salts from China	7–680 particles/kg	Contamination by microplastic particles	[45]
Sugar	217 ± 124 fibers/kg; 33±7 fragments/kg	Contamination by transparent and color particles and fibers	[39]
Canned sardines and sprats	1–3 particles/can	Contamination by plastic polymers	[46]
Beer from German supermarket	2–79 fibers/L; 12–109 fragments/L	Contamination by fibers, fragments and granules	[47]
Bottled water	10.4–325 particles	Contamination by particulate fragments and fibers	[60]
Single-use and reusable bottles	Single use: 14 particles/L; Reusable bottles: 118 particles/L	Contamination by fragmented particles of polypropylene, polyethylene tetraphthalate	[48]
Tap water	0–61 particles/L	Contamination by fibers	[43]

made available from different countries [39,40]. Cellulose fibers and chitin fragments were found as particulate matter in honey. In the second of the two studies by Liebezeit, the honey samples from Germany were found to contain microplastics in the range of 40–660 fibers/kg of honey and 0–38 fragments/kg of honey [39,40,47]. However, the authors did not classify the fragments. Mühlschlegel et al. studied honey from Switzerland for possible contamination by fragmented particles; they also suggested possible precautions for minimizing them [41].

Contamination of salts is directly related to the microplastics in aquatic environments. Karami et al., in their studies, confirmed the presence of significant amounts of micro- and nanofragmented plastics in salt samples. Karami et al., also studied 17 different brands of salts available in a Malaysian market and concluded the presence of microplastics by Raman spectroscopy [44]. In another study by Gündoğdu, microplastics were confirmed in Turkish salt by μ-Raman spectroscopy [42]. Kosuth et al. studied commercialized salt samples made available from US grocery stores and confirmed the presence of significant amounts of microplastics [43]. Yang and co-workers studied sea, rock and lake salt from China and concluded the presence of 7–680 particles/kg [45]. One study by Liebezeit and Liebezeit reported possible contamination of sugar by microplastics. They concluded that unrefined sugar had the highest number of fragmented particles and fibers [39]. Karami et al., also studied canned sprat and sardine brands made available from 13 different countries. By exploiting Raman spectroscopy, the researchers concluded an average presence of 1–3 particles in four out of 20 brands [46]. Liebezeit et al. also studied and confirmed the possible presence of microplastics in German beers [47]. A similar kind of study was conducted with beers from the United States by Kosuth et al. and confirmed the presence of microplastics in US beers [43].

Mason et al. studied 259 bottled waters from nine different countries for possible contamination by micro- and nanoplastics. It was also concluded that most of the contamination would be from packaging sources [59,60]. In a similar study by Schymanski et al., 22 single-use bottled waters were compared with reusable plastic bottles for contamination with microplastics. The authors concluded 8.4 times greater contamination in reused bottles than single-use bottles [48]. Kosuth et al. studied tap water collected from 14 different countries and concluded that the average presence of microplastics was 0–61 particles/L [43].

Packaging materials usually contain chemical constituents, additives, monomers, etc. which may, upon contact with food, leach into the food product [61,62]. However, such contamination is not a spontaneous process. Currently nanotechnology is being applied to almost every scientific and nonscientific field. Nanoparticles are now used for designing food packaging that can improve the shelf life and freshness of the product [63]. There is a strong possibility of migration of such nanoparticles from packaging material to food upon contact [64].

6.6 HEALTH HAZARDS CAUSED BY FRAGMENTED PLASTICS

Humans are consuming a huge amount of micro- and nanoplastics through ingestion and inhalation. These fragmented plastics are capable of accumulating and exerting chemical toxicity and induce or enhance immune response. Chemical toxicity usually occurs due to localized leaching of additives or adsorbed pollutants. Chronic exposure to such fragmented plastics and chemicals is of greater concern.

6.6.1 Toxicological Pathways of Fragmented Plastics

Microplastics are capable of generating toxicological effects upon consumption by the oral or inhalational route. Solubility, size, shape and surface charge collectively influence the cytotoxic effect of particles to tissues *in vivo* [65]. If microplastics are retained in the biological system, they may lead to genotoxicity, apoptosis, oxidative stress and even necrosis of cells. Upon prolonged exposure, these conditions may even worsen to cause carcinogenesis, fibrosis and damage to tissues. Oxidative stress and inflammation have been reported upon inhalation of micro- and nanoplastics [65]. Micro- and nanoplastics, upon weathering, form free radicals by dissociation of the C–H bonds [66,67]. These free radicals pose more danger to human health than the microplastics.

6.6.2 Inflammatory and Immune Response by Fragmented Plastics

Extensive research is available concerning inflammation due to microparticles formed owing to abrasion of prosthetic implants made up of plastics. Individuals with plastic endoprostheses have been reported to contain microparticles of varying shapes in the joint cavity and joint capsules [68]. Particles of less than 1 μm accumulate in the mobile macrophages by means of lymphatic transport [69]. These particulate matters provoke immune activation of macrophages and production of cytokines [70]. All these results suggest that the chemical composition of micro- and nanoplastics determines the type of immunological response upon exposure.

6.6.3 Effect on the Airway and Gastrointestinal Tract

Macrophages along with antigens and toxins, in conjugation with microparticles, enhance T-cell proliferation [71]. Corona formed on microplastics not only influences particle uptake but also induces toxicity [72,73]. Zeta potential, surface charge, size and shape also tend to influence toxicity. In a study on rats, smaller particles have been found to induce greater influx to neutrophils and cause inflammation of lungs [74]. There are limited data available on inflammatory responses in the gastrointestinal tract.

6.6.4 Effect of Adsorbed and Endogenous Chemical Pollutants

Fragmented particles with large surface area-to-volume ratios of microplastics and hydrophobicity due to adsorbed hydrocarbons on the surface cause toxicity in humans. Some of the adsorbed hydrocarbons have been found to be carcinogenic, mutagenic and immunotoxic in nature. The additives mixed while manufacturing of plastics can leach out based on concentration gradient [75]. Upon accumulation, microplastics present a new source of chemicals to tissues and body fluids. They are capable of inducing reproductive toxicity, mutagenicity, carcinogenicity and even disruption of hormones [76]. Inhalation and ingestion of household dust is a predominant source of micro- and nanoplastics. Carpets, electronics and upholstery contain polybrominated diphenyl ethers (PBDEs), which are also toxic in nature. Apart from chemical additives, microplastics also contain unreacted monomers, which are classified as mutagenic, carcinogenic or both [77]. In a recent study on mice, it was found that peritoneal dialysis solution contains leached contaminants [78].

6.6.5 Effect on Microbiome

Microbiome refers to the microbial community living inside or on the body and assisting the host in its physiological well-being [79]. If the microbiomes colonize over microplastics, the balanced composition of the microbes will be hampered, and thus it will impair the physiological functions. Microbiomes have the capacity of metabolizing various environmental toxicants, thus forming colonies over microplastics may affect immunogenic responses [79]. Lungs also host a number of microbial communities; colonization of microbes in gastrointestinal and airway tracts may alter immunogenic responses in such areas. However, colonization has been found to be lower in the respiratory region relative to gastrointestinal region. Thus, colonization may cause a change in the microbial composition in the gastrointestinal tract and lungs, which may lead to oxidative stress.

6.7 CURRENT METHODS OF DETECTION, IDENTIFICATION AND QUANTIFICATION OF MICRO- AND NANOPLASTICS IN FOOD AND OTHER SAMPLES

One of the major challenges for the detection and analysis of micro- and nanofragmented plastics is the lack of a standard operating procedure and quality assurance. Apart from that, there is no detailed and internationally agreed definition of micro- and nanoplastics. Currently, both conventional and advanced detection or identification methods are being utilized by scientists globally. In this chapter, a brief description of measurement principle and advantages and disadvantages of commonly used methods of detection have been summarized.

Further information may be found in Chapters 7 and 8 of this book.

6.7.1 Filtration Technique

It is one of the most basic methods for recovering microplastics from liquids like water and other solutions. The process is simple, cheap, categorizes particles according to size and is suitable for liquids. However, it comes with the limitation of clogging of pores of the filter, loss of particles with large pore size filters, and inaccurate data.

6.7.2 Floatation or Sedimentation Technique

Differences in the densities of inorganic materials present in microplastics is exploited to extract particulates of plastic by floating them out from salt solution, whose density is higher than plastic but lower than other solids and minerals present in them [80]. Particles can be made to sediment or float depending on their size, or density, because of the effect of gravitational force. The process can be accelerated by centrifugation. It separates microplastics from inorganic matter. However because the plastic particles need to be detached from the matrix material, the critical choice of the type of gradient is important.

6.7.3 Matrix Dissolution Technique

This technique is used to study the traces of fragmented plastics in animal tissues, organs and biofilms. The acid digestion technique used for the detection of trace elements has also been exploited for microplastics [81]. This method needs optimization to digest organic matter. There is a possibility that the plastic particles may get damaged or destroyed by acids. Alkaline digestion by NaOH or KOH has been reported for the detection of microplastics in biological matrices [82]. The process has been successfully used for a wide range of polymers because of the low risk of damaging or chemically modifying particulates. Biological materials are usually detected by the enzyme digestion technique. However, certain enzymes are specific to certain types of protein matrices, and thus it requires several sequential steps to remove the matrix components adequately.

6.7.4 Oil Extraction

Plastics are hydrophobic in nature and they can be extracted in a purely aqueous medium. Crichton et al., in their study, exploited the oleophilic property of fragmented plastics to separate them from aqueous medium by partitioning them into water-immiscible oil [83]. This method only separated microplastics from their inorganic matter and is compatible with FTIR. This process may cause damage to low melting point polymers.

6.7.5 Optical Vibrational Spectroscopy

This technique is based on irradiating the sample to generate molecular vibration, producing an optical spectrum. Based on the particular peaks of the spectrum, identification of the material is done. This method provides reproducible results and requires only a very small quantity of samples. Also, the process is non-intrusive and nondestructive in nature. Detection of microplastics is usually done by ultraviolet-induced photodegradation. FTIR (Fourier-transform infrared) spectroscopy and Raman spectroscopy are also used. Raman spectroscopy is a laser-based technique which provides better resolution than IR spectroscopy. They are capable of identifying microplastics of about a 1 μm size [48,84]. This technique is widely used for detection of particulate matter in combination with FTIR for qualitative analysis, detection and characterization.

6.7.6 Thermal Analysis

Thermal analysis is usually used in combination with gas chromatography mass spectrometric technique (GC-MS). This method is destructive in nature and provides no, or limited, data about the size and shape of the particulate matter.

6.7.7 Fluorescent Tagging with Nile Red

Nile red has been exploited to fluorescently label microplastics upon irradiation with blue light. This method can quantify microplastics of 20 μm–1 mm [85–87]. The dye is relatively cheap and readily available. Erni-Cassola et al., Maes et al. and Shim et al. have reported tagging of the microplastics with Nile red and subsequently applying density separation and filtration of particles on black polycarbonate paper [85–87].

6.7.8 Time-of-Flight Secondary Ion Mass Spectrometry (ToF-SIMS)

This technique provides ambient information about the chemical composition of particulate solid samples upon bombarding the sample surface with a focused pulse ion beam. After that, the material is analyzed under a time-of flight mass spectrometer, which provides information about the elemental and molecular composition of the sample. Microplastics can also be analyzed by this technique [88]. However, it is an expensive technique with complex instrumentation.

6.8 CONCLUSION AND FUTURE PROSPECTIVE

Micro- and nanoplastics are impacting life of humans and other living creatures. Once a boon, the overproduction and waste of plastics have now become the worst environmental issue of all time. The microfragmented plastics reach the human diet through trophic level transfer and bioaccumulation. There are massive litters of plastics found almost in every country; still, we know little about the toxicological effects of micro- and nanoplastics on humans. This is due to limited data availability.

Pertaining to the varying variety of particle size, composition, shape and their ability to adsorb other pollutants, there are very limited methods that can be easily used for analysis purpose. There is also lack of standard materials to be used as reference. In the current scenario, only extrapolation of small sample size data can be done. However, it is possible that it may result in inaccurate data. Thus, the management of this invisible danger will become more complex day by day.

REFERENCES

1. Mitrano D. Nanoplastic should be better understood. *Nature Nanotechnol* 2019, 14, 299.
2. UN Environmental Annual Report. 2017. https://www.unenvironment.org/annualreport/2017/ (Accessed on 30-09-2019)
3. Rocha-Santos T and Duarte AC. A critical overview of the analytical approaches to the occurrence, the fate and the behavior of microplastics in the environment. *Trends in Anal Chem* 2015, 65, 47–253.
4. Jambeck JR, Geyer R, Wilcox C, Siegler TR, Perryman M, Andrady A, Narayan R, and Law KL. Plastic waste inputs from land into the ocean. *Science*. 2015, 347(6223), 768–771.
5. EFSA. Presence of microplastics and nanoplastics in food, with particular focus on seafood. EFSA Panel on Contaminants in the Food Chain (CONTAM). *EFSA Journal* 2016, 14(6), 4501.
6. GESAMP. Sources, fate and effects of microplastics in the marine environment: Part two of a global assessment. In: Kershaw PJ, and Rochman CM (eds) *(IMO/FAO/UNESCO-IOC/UNIDO/WMO/IAEA/UN/UNEP/UNDP Joint Group of Experts on the Scientific Aspects of Marine Environmental Protection)*. GESAMP Reports & Studies No 93, 2016.
7. Eu (European Union). Report for Europena Commission DG Environment. Study to support the development of measures to combat a range of marine litter sources. Report for European Commission DG Environment, 2016.
8. Bouwmeester H, Hollman PCH, Peters RJB Potential health impact of environmentally released micro- and nanoplastics in the human food production chain: Experiences from nanotoxicology. *Environ Sci Technol*. 2015, 49, 8923–8947.
9. Fries E, Dekiff JH, Willmeyer J, Nuelle M-T, Ebert M, Remy D. Indentification of polymer types and additives in amrine microplastic particles using prolysis-GC/MS and scanning electron microscopy. *Environ Sci. Process Impacts* 2013, 15, 1949–1956.

10. Andrady AL. Microplastics in the marine environment. *Mar Pollut Bull*. 2011, 62, 1596–1605.

11. Koelmans AA, Besseling E, Shim WJ. Nanoplastics in the aquatic environment. Critical review. In: Bergmann M, Gutow L, and kLages M (eds). *Marine Anthropogenic Litter*. Springer International Publishing, Cham, 2015, pp. 325–340.

12. Lambert S and Wagner M. Characterisation of nanoplastics during the degradation of polystyrene. *Chemosphere*. 2016, 145, 265–268.

13. Zettler ER, Mincer TJ, Amaral-Zetter LA. Life in the plstispher: Microbial communities on plastic marine debris. *Environ Sci Technol*. 2013, 47, 7137–7146.

14. Cozar A, Echevarría F, González-Gordillo JI, Irigoien X, Ubeda B, Hernández-León S, Palma AT et al. Plastic debris in the open ocean. *Proc the Nat Acad Sci the USA* 2014, 111(28), 10239–10244.

15. Lusher A, Hollman P, and Medonza-Hill Microplastics in Fisheries and Qquaculture. Status of knowledge on their occurrence and implications for aquatic organisms and food safety. FAO Fisheries and Aquaculture Technical Paper No 615. FAO, Rome, 2017.

16. Jovanović B. Ingestion of microplastics by fish and its potential consequences from a physical perspective. *Integr Environ Assess Manag*. 2017 May, 13(3), 510–515.

17. Mazurais D, Ernande B, Quazuguel P, Severe A, Huelvan C, Madec L, Mouchel O et al. Evaluation of the impact of polyethylene microbeads ingestion in European sea bass (Dicentrarchuslabrax) larvae. *Marine Environ Res* 2015, 112, 78–85.

18. Peda C, Caccamo L, Fossi MC et al. Intestinal alterations in European sea bass Dicentrarchuslabrax (Linnaeus, 1758) exposed to microplastics: Preliminary results. *Environ Pollut*. 2016, 212, 251–256.

19. Cedervall T, Hansson LA, Lard M, Frohm B, Linse S. Food chain transport of nanoparticles affects behaviour and fat metabolism in fish. *PLOS ONE* 2012, 7, e32254.

20. Lu Y, Zhang Y, Deng Y. Uptake and accumulation of polystyrene microplastics in zebrafish (Danioi rerio) and doxic effects in liver. *Environ Sci Technol*. 2016, 50, 4054–4060.

21. Rochman CM, Hoh E, Kurobe T, Teh SJ. Ingested plastic tranfers hazardous chemicals to fish and induces hepatic stress. *Sci Rep*. 2013, 3, 3263.

22. Wright SL and Kelly FJ. Plastic and human health: A micro issue?. *Environmental Sci Technol* 2017 Jun 7, 51(12), 6634–6647.

23. Kasirajan S and Ngouajio M. Polyethylene and biodegradable mulches for agricultural 913 applications: A review. *Agron Sustain Dev* 2012, 32(2), 501–529.

24. Dris R, Gasperi J, Saad M, Mirande C, and Tassin B. Synthetic fibers in atmospheric fallout: A source of microplastics in the environment? *Mar. Pollut. Bull* 2016, 104(1–2), 290–293.

25. Warheit DB, Hart GA, Hesterberg TW, Collins JJ, Dyer WM, Swaen GMH, Castranova V, Soiefer AI, and Kennedy GL Jr. Potential pulmonary effects of man-made organic fiber (MMOF) dusts. *Crit Rev Toxicol*. 2001, 31(6), 697–736.

26. Boag AH, Colby TV, Fraire AE, Kuhn C 3rd, Roggli VL, Travis WD, and Vallyathan V. The pathology of interstitial lung disease in nylon flock workers. *Am J Surg Pathol*. 1999, 23(12), 1539–1545.

27. Eschenbacher WL, Kreiss K, Lougheed MD, Pransky GS, Day B, and Castellan RM. Nylon flock associated interstitial lung disease. *Am J Respir Crit Care Med*. 1999, 159(6), 2003–2008.

28. Pimentel JC, Avila R, and Lourenço AG. Respiratory disease caused by syntheticfibres: A new occupational disease. *Thorax*. 1975, 30(2), 204–219.

29. Pauly JL, Stegmeier SJ, Allaart HA, Cheney RT, Zhang PJ, Mayer AG, and Streck RJ. Inhaled cellulosic and plastic fibers found in human lung tissue. *Cancer Epidemiol. Biomarkers Prev* 1998, 7 (5), 419–428.

30. Geiser M and Kreyling WG. Deposition and biokinetics of inhaled nanoparticles. *Part Fibre Toxicol* 2010, 7(2), doi: 10.1186/1743-8977-7-2.

31. Deville S, Penjweini R, Smisdom N, Notelaers K, Nelissen I, Hooyberghs J, and Ameloot M. Intracellular dynamics and fate of polystyrene nanoparticles in A549 Lung epithelial cells monitored by image (cross-) correlation spectroscopy and single particle tracking. *Biochim. Biophys. Acta* 2015, 1853(10, Part A), 2411–2419.

32. Geiser M, Rothen-Rutishauser B, Kapp N, Schurch S, Kreyling W, Schulz H, Semmler M, Im Hof V, Heyder J, and Gehr P. Ultrafine particles cross cellular membranes by nonphagocytic mechanisms in lungs and in cultured cells. *Environ Health Perspect.* 2005, 113(11), 1555–1560.

33. Powell JJ, Faria N, Thomas-McKay E, and Pele LC. Origin and fate of dietary nanoparticles and microparticles in the gastrointestinal tract. *J Autoimmun.* 2010, 34(3), J226–233.

34. Sass W, Dreyer HP, and Seifert J. Rapid insorption of small particles in the gut. *Am J Gastroenterol.* 1990, 85(3), 255–260.

35. Volkheimer G. The phenomenom of persorption: Persorption, dissemination, and elimination of microparticles. In: Heidt PJ, Nieuwenhuis P, Rusch VD, and Waaij DVD (eds) *Old Herborn University Seminar Monograph* Vol. 14. Germany, 2001, pp. 7–13.

36. Freedman BJ. Persorption of raw starch: A cause of senile dementia? *Med. Hypotheses* 1991, 35(2), 85–87.

37. Volkheimer G. Hematogenous dissemination of ingested polyvinyl chloride particles. *Ann NY Acad Sci* 1975, 246(1), 164–171.

38. Volkheimer G and Schulz FH. The phenomenon of persorption. *Digestion.* 1968, 1(4), 213–218.

39. Liebezeit G and Liebezeit E. Non-pollen particulates in honey and sugar. *Food Addit Contam Part A Chem Anal Control Expo Risk Assess* 2013, 30(12), 2136–2140.

40. Liebezeit G and Liebezeit E. Origin of synthetic particles in honeys. *Pol J Food Nutr Sci.* 2015, 65(2), 143–147. doi:10.1515/ pjfns-2015–0025.

41. Mühlschlegel P, Hauk A, Walter U, and Sieber R. Lack of evidence for microplastic contamination in honey. *Food AdditContam Part A Chem Anal Control Expo Risk Assess* 2017, 34(11), 1982–1989.

42. Gündoğdu S. Contamination of table salts from Turkey with microplastics. *Food Addit Contam Part A Chem Anal Control Expo Risk Assess* 2018, 35(5), 1006–1014.

43. Kosuth M, Mason SA, and Wattenberg EV. Anthropogenic contamination of tap water, beer, and sea salt. *PLOS ONE* 2018, 13(4), e0194970.

44. Karami A, Golieskardi A, Ho YB, Larat V, and Salamatinia B. Microplastics in eviscerated flesh and excised organs of dried fish. *Nat Sci Rep* 2017, 7(5473), 2045–2322.

45. Yang D, Shi H, Li L, Li J, Jabeen K, and Kolandhasamy P. Microplastic pollutionin table salts from China. *Environ Sci Technol.* 2015, 49(22), 13622–13627.

46. Karami A, Golieskardi A, Keong Choo C, Larat V, Karbalaei S, and Salamatinia B. Microplastic and mesoplastic contamination in canned sardines and sprats. *Sci Total Environ.* 2018, 612, 1380–1386.

47. Liebezeit G and Liebezeit E. Synthetic particles as contaminants in German beers. *Food AdditContam Part A Chem Anal Control Expo Risk Assess* 2014, 31(9), 1574–1578.

48. Schymanski D, Goldbeck C, Humpf H-U, and Fürs P. Analysis of microplastics in water by micro-Raman spectroscopy: Release of plastic particles from different packaging into mineral water. *Water Res.* 2018, 129, 154–162.
49. Carbery M, O'Connor W, and Thavamani P. Trophic transfer of microplastics and mixed contaminatns in the marine food web and implications for human health. *Environ Int.* 2018, 115, 400–409.
50. Huerta Lwanga E, Mendoza Vega J, Ku Quej V, de Los Angeles Chi J, Sanchez Del Cid L, Chi C, Escalona Segura G et al. Field evidence for transfer of plastic debris along a terrestrial food chain. *Nat Sci Rep* 2017, 7, 14071.
51. Boerger CM, Lattin GL, Moore SL, and Moore CJ. Plastic ingestion by planktivorous fishes in the North Pacific Central Gyre. *Mar Pollut Bull.* 2010, 60(12), 2275–2278.
52. Possatto FE, Barletta M, Costa MF, Ivar Do Sul JA, and Dantas DV. Plastic debris ingestion by marine catfish: An unexpected fisheries impact. *Mar Pollut Bull.* 2011, 62(5):1098–1102.
53. Thompson RC, Olsen Y, Mitchell RP, Davis A, Rowland SJ, John AWG, McGonigle D, and Russel AE. Lost at sea: Where is all the plastic? *Science.* 2004, 304(5672), 838.
54. Hollman PCH, Bouwmeester H, and Peters RJB. Microplastics in the aquatic food chain; Sources, measurement, occurrence and potential health risks. RIKILT report 2013.003. RIKILT Wageningen UR (University & Research Centre), Wageningen, 2013.
55. Browne MA, Dissanayake A, Galloway TS, Lowe DM, and Thompson RC. Ingested microscopic plastic translocates to the circulatory system of the mussel, Mytilus edulis (L.). *Environ Sci Technol.* 2008, 42(13), 5026–5031.
56. Li J, Green C, Reynolds A, Shia H, and Rotchell JM. Microplastics in mussels sampled from coastal waters and supermarkets in the United Kingdom. *Environ Pollut.* 2018, 241, 35–44.
57. Li J, Yang D, Li L, Jabeen K, and Shi H. Microplastics in commercial bivalves from China. *Environ Pollut.* 2015, 207, 190–195.
58. Li WC, Tse HF, and Fok L. Plastic waste in the marine environment: A review of sources, occurrence and effects. *Sci Total Environ.* 2016, 566–567, 333–349. doi:10.1016/j. scitotenv.2016.05.084
59. Mason SA, Garneau D, Sutton R, Chu Y, Ehmann K, Barnes J, Fink P, Papazissimos D, and Rogers DL. Microplastic pollution is widely detected in US municipal wastewater treatment plant effluent. *Environ Pollut.* 2016, 218, 1045–1054.
60. Mason SA, Welch V, and Neratko J. Synthetic polymer contamination in bottled water. *Front Chem.* 2017, 6, 407. doi:10.3389/fchem.2018.00407.
61. Castle L. Chemical migration into food: An overview. In: Barnes K, Sinclair R, and Watson D (eds) *Chemical Migration and Food Contact Materials.* Woodhead Publishing, Sawston (Cambridge), 2007, pp. 1–13.
62. Cooper I. Plastics and chemical migration into food. In: Barnes K, Sinclair R, and Watson D (eds) *Chemical Migration and Food Contact Materials.* Woodhead Publishing, Sawston (Cambridge), 2007, pp. 228–250.
63. Bumbudsanpharoke N, Choi J, and Ko S. Applications of nanomaterials in food packaging. *J Nanosci. Nanotechnol* 2015, 15(9), 6357–6372.
64. Addo Ntim S, Norris S, Goodwin DG Jr., Breffke J, Scott K, Sung L, Thomas TA, and Noonan GO. Effects of consumer use practices on nanosilver release from commercially available food contact materials. *Food Addit Contam Part A Chem Anal Control Expo Risk Assess* 2018, 35(11), 2279–12290.

65. Nel A, Xia T, Mädler L, and Li N. Toxic potential of materials at the nanolevel. *Science*. 2006, 311(5761), 622–627.

66. White J and Turnbull A. Weathering of polymers: Mechanisms of degradation and stabilization, testing strategies and modelling. *J Mater Sci*. 1994, 29(3), 584–613.

67. Gewert B, Plassmann MM, and MacLeod M. Pathways for degradation of plastic polymers floating in the marine environment. *Environ Sci.: Process Impacts* 2015, 17, 1513.

68. Willert HG and Semlitsch M. Tissue reactions to plastic and metallic wear products of joint endoprostheses. *Clin. Orthop. Relat. Res* 1996, 333, 4–14.

69. Urban RM, Jacobs JJ, Tomlinson MJ, Gavrilovic J, Black J, and Peoc'h, M. Dissemination of wear particles to the liver, spleen, and abdominal lymph nodes of patients with hip or knee replacement. *J Bone Joint Surg* 2000, 82(4), 457–457.

70. Hicks DG, Judkins AR, Sickel JZ, Rosier RN, Puzas JE, and O'Keefe RJ. Granular histiocytosis of pelvic lymph nodes following total hip arthroplasty. The presence of wear debris, cytokine production, and immunologically activated macrophages. *J Bone Jt Surg* 1996, 78(4), 482–496.

71. Kovacsovics-Bankowski M, Clark K, Benacerraf B, and Rock KL. Efficient major histocompatibility complex class I presentation of exogenous antigen upon phagocytosis by macrophages. *Proc Nat Acad Sci USA* 1993, 90(11), 4942–4946.

72. Evans SM, Ashwood P, Warley A, Berisha F, Thompson RPH, and Powell JJ. The role of dietary microparticles and calcium in apoptosis and interleukin-1β release of intestinal macrophages. *Gastroenterology*. 2002, 123(5), 1543–1553.

73. Lundqvist M, Stigler J, Elia G, Lynch I, Cedervall T, and Dawson KA. Nanoparticle size and surface properties determine the protein corona with possible implications for biological impacts. *Proc Nat Acad Sci USA* 2008, 105(38), 14265–14270.

74. Brown DM, Wilson MR, MacNee W, Stone V, and Donaldson K. Size-dependent proinflammatory effects of ultrafine polystyrene particles: A role for surface area and oxidative stress in the enhanced activity of ultrafines. *Toxicol Appl Pharmacol*. 2001, 175(3), 191–199.

75. Tickner J. *The use of Di-2-Ethylhexyl Phthalate in PVC Medical Devices: Exposure, Toxicity, and Alternatives*. Lowell Centre for Sustainable Production, 1999.

76. Mariana M, Feiteiro J, Verde I, and Cairrao E. The effects of phthalates in the cardiovascular and reproductive systems: A review. *Environ Int*. 2016, 94, 758–776.

77. Lithner D, Larsson A, and Dave G. Environmental and health hazard ranking and assessment of plastic polymers based on chemical composition. *Sci Total Environ*. 2011, 409(18), 3309–3324.

78. Gader Al-Khatim ASA, and Galil KAA. Postnatal toxicity in mice attributable to plastic leachables in peritoneal dialysis solution (PDS). *Arch Environ Occup Health* 2015, 70(2), 91–97.

79. Cho I and Blaser MJ. The human microbiome: At the interface of health and disease. *Nat Rev Genet*. 2012, 13(4), 260–270.

80. Quinn B, Murphy F, and Ewins C. Validation of density separation for the rapid recovery of microplastics from sediment. *Anal Methods* 2017, 9, 1491–1498.

81. Miller ME, Kroon FJ, and Motti CA. Recovering microplastics from marine samples: A review of current practices. *Mar Pollut Bull*. 2017, 123(1), 6–18.

82. Enders K, Lenz R, Beer S, and Stedmon CA. Extraction of microplastic from biota: Recommended acidic digestion destroys common plastic polymers. *ICES J Mar Sci*. 2016, 74(1), 326–331.

83. Crichton EM, Noël M, Giesab EA, and Ross PS. A novel, density-independent and FTIR-compatible approach for the rapid extraction of microplastics from aquatic sediments. *Anal Methods* 2017, 9(9), 1419–1428.

84. Rocha-Santos TAP, and Duarte AC. *Comprehensive Analytical Chemistry. Characterization and Analysis of Microplastics.* Vol. 75, 1st. Elsevier, Oxford, 2017.

85. Erni-Cassola G, Gibson M, Thompson RC, and Christie-Oleza JA. Lost, but found with nile red: A novel method for detecting and quantifying small microplastics (1 mm to 20 μm) in environmental samples. *Environ Sci Technol.* 2017, 51(23), 13641–13648.

86. Maes T, Jessop R, Wellner N, Haupt K, and Mayes AG. A rapid-screening approach to detect and quantify microplastics based on fluorescent tagging with Nile Red. *Nat Sci Rep* 2017, 7, 44501.

87. Shim WJ, Hong SH, and Eo SE. Identification methods in microplastic analysis: A review. *Anal Methods* 2017, 9(9), 1384–41391.

88. Jungnickel H, Pund R, Tentschert J, Reichardt P, Laux P, Harbach H, and Luch A. Time-of-flight secondary ion mass spectrometry (ToF-SIMS)-based analysis and imaging of polyethylene microplastics formation during sea surf simulation. *Sci Total Environ.* 2016, 563–564, 261–266.

89. Toussaint B, Raffael B, Angers-Loustau A, Gilliland D, Kestens V, Petrillo M, Rio-Echevarria I, and Van den Eede G. Review of micro- and nanoplastic contamination in the food chain. *Food Additives Contaminants: Part A*, 2019, 36(5), 639–673.

90. Wright SL, Thompson RC, Galloway TS. The physical impacts of microplastics on marine organisms: *A review. Environ. Pollut.*, 2013, 178, 483–492.

SECTION IV

Sampling and Analysis

CHAPTER 7

Sampling, Isolating and Digesting of Microplastics

Leo M. L. Nollet

CONTENTS

A. L. Lusher et al. [1] published an interesting and valuable article on sampling, isolating and identifying microplastics ingested by fish and invertebrates.

7.1 SAMPLING

7.1.1 Field-Collected Organisms

Micro- and nanoplastics are taken up by a wide range of organisms in a diverse range of habitats, including the sea surface, water column, benthos, estuaries, beaches and aquaculture [2]. The diversity of the organisms and habitats where they live and are sampled require a range of collection techniques [3–5]. The sampling method is determined by the research question, available resources, habitat and target organism. Benthic invertebrate species such as *Nephrops norvegicus* may be collected in grabs, traps and creels, or by bottom trawling [6,7]. Planktonic and nektonic invertebrates are collected by way of manta and bongo nets [8–11]. Fish species are generally recovered in surface, midwater and benthic trawls, depending on their habitats. Gill nets have been used in riverine systems [12]. Some species are collected from the field by hand; this is common practice for bivalves, crustaceans and annelids [13–17]. Another method is direct collection from shellfish or fish farms [18–20] or from commercial fish markets, where the capture method is often unknown [21–22].

7.1.1.1 Microplastic Losses during Field Sampling

Handling stress, physical movement and the physiological and behavioral specificities of the sampled organism may result in the loss of microplastics prior to animal preservation. Gut evacuation times for animals range from minutes for decapod crustaceans to several hours for calanoid copepods [10] and fish [23,24] to days in larger lobsters [25]. Some animals might egest microplastic debris prior to analysis [7]. In such cases, the time between sample collection and the preservation of the animal must be as short as possible. Care must also be taken to minimize handling stress or physical damage.

The copepod *Eurytemora affinis* [26] and some fish species have been observed regurgitating their stomach contents [27]. Compression of a catch in the cod end might induce regurgitation in fish [29]. The likelihood of regurgitation increases with depth of capture, and gadoids are more prone to regurgitation than flatfish. Piscivorous predators are prone to regurgitation owing to their large distensive esophagus and stomach [28, 30].

7.1.1.2 Microplastic Accumulation during Field Sampling

Laboratory studies have identified that nano- and microplastics can adhere to external appendages of marine copepods [10]. Cataloguing such interactions in nature is complicated as determining whether the resulting accumulation has occurred naturally or as a by-product of the sampling regimen is difficult. A similar interaction may occur with organisms feeding on microplastics during capture in nets; this is particularly of concern when the mesh size of the net is capable of collecting microplastics, for example, in manta nets (common mesh size 0.33 mm) [23].

7.1.1.3 Sample Storage

It is important how biotic samples are stored. The choice of the preservation technique will largely depend on the following analysis. It is important to know if the fixative will affect the structure, microbial surface communities, chemical composition, color or analytical properties of microplastics within the sample. Four percent formaldehyde and 70% ethanol are commonly used fixatives; however, these preservatives at higher concentrations can damage some polymers. Polyamide is only partially resistant to 10% formaldehyde solution, while polystyrene can be damaged by 100% alcohol. Alternative methods for storage of organisms include desiccation [9] and freezing [7,31–33].

7.1.2 Laboratory-Exposed Organisms

Laboratory studies are used to better understand the interactions between microplastics and biota. Controlled laboratory exposures facilitate monitoring of the uptake, movement and distribution of synthetic particles in whole organisms and excised tissues. Fluorescently labeled plastics, either purchased or dyed in the lab [34], allow visualization of microplastics in organisms with transparent carapaces, [10,18,35] circulatory fluids, (36,37) or histological sections [38]. Where dissection is prohibitive (e.g., mussels) fluorescent microplastics can be quantified by physically homogenizing tissues followed by microscopic analysis of sub-sampled homogenate [14]. Coherent anti-Stokes Raman scattering (CARS) has also been used to visualize non-fluorescent nano- and microplastics in intestinal tracts and those adhered to external appendages of copepods and gill lamellae of crabs [10,14]. Bioimaging techniques, however, are not feasible, with field sampled biota as environmental plastics do not fluoresce and may be obscured by tissues or algal fluorescence.

7.2 ISOLATING MICROPLASTICS

In recent years, many techniques have been developed to detect microplastics in biota. Methods for extracting microplastics from biotic material include dissection, depuration, homogenization and digestion of tissues with chemicals or enzymes.

7.2.1 Dissection

In larger animals, including squid [39], whales [40–41], turtles [42] and seabirds [43], dissection of the gastrointestinal tract and subsequent quantification of synthetic particles from the gut is the method of choice for assessing plastic consumption. In laboratory studies, it is more common for the whole organism (42% of studies) or the digestive tract (26% of studies) to be digested or analyzed. In comparison, 69% of field studies targeted the digestive tract, and 27% looked at the whole organism. In invertebrates and vertebrates and in pelagic and demersal fish, the excision of the intestinal tract can also be used to look for microplastics [6,7,23,32,44–68]. Investigation of stomachs and intestines is relevant for microplastics >0.5 mm in size. Microplastics larger than this do not readily pass through the gut wall without pre-existing damage, and the likelihood of translocation into tissues is too low to warrant regular investigation [69,70]. Localization of microplastics <0.5 mm can be determined by excising organs, such as the liver or gills [38,71,72], or, where the research question relates to risks of human consumption, edible tissues, for example, tail muscles of shrimp [73]. Microplastics present in dissected tissues can be isolated using saline washes, density flotation, visual inspection, or digestion.

7.2.2 Depuration

Any externally adhered plastics can be removed prior to treatment by washing the study organism with water or saline water or using forceps [11,74]. A depuration step can be used to eliminate transient microplastics present in the intestinal tract. Depuration is facilitated by housing animals in microplastic-absent media (e.g., freshwater, seawater, sediment), with or without food, and leaving sufficient time for complete gut evacuation [75]. These media should be refreshed regularly to prevent consumption of egested microplastics [76]. Depuration ensures that only microplastics retained within tissues or entrapped in the intestinal tract are considered [15,76,78]. Depuration makes it also feasible to collect fecal matter, typically sampled via siphon, sieve or pipette. These feces can subsequently be digested, homogenized or directly visualized to assess and quantify egested microplastics. Fecal analysis has been used to determine microplastic consumption in a range of taxa, including sea cucumbers [79], copepods [34,80], isopods [16], amphipods [81], polychaetes [95] and mollusks [82–94].

7.3 DIGESTION

Enumerating microplastics in biota, excised tissues or environmental samples can be challenging because the plastic may be masked by biological material, microbial biofilms, algae and detritus [9]. To isolate microplastics, organic matter can be digested, leaving only recalcitrant materials (Table 7.1). Traditionally, digestion is conducted using strong oxidizing agents. Synthetic polymers, however, can be degraded or damaged by these chemical treatments, particularly at higher temperatures. Environmentally exposed plastics, which may have been subject to weathering, abrasion and photodegradation, may have reduced structural integrity and resistance to chemicals compared to that of virgin plastics used in these stress tests [96]. As such, data ascertained using caustic digestive agents should be interpreted with caution, and the likely loss of plastics from the digestive treatment carefully considered.

TABLE 7.1   Optimized Protocols for Digesting Biota or Biogenic Material to Isolate Microplastics

Treatment	Exposure	Organism	Reference
HNO_3 (22.5 M)	20°C (12 h) + 100°C (2 h)	Blue mussels	78
HNO_3 (22.5 M)	20°C (12 h) + 100°C (2 h)	Blue mussels oysters	75
HNO_3 (22.5 M)	20°C (12 h) + 100°C (2 h)	Blue mussels lugworms	76
HNO_3 (100%)	20°C (30 min)	Euphausids copepods	11
HNO_3 (69%–71%)	90 °C (4 h)	Manilla clams	74
HNO_3 (70%)	2 h	Zebrafish	38
HNO_3 (22.5 M)	20°C (12 h) + 100°C (15 min)	Brown mussels	104
HNO_3 (65%) $HClO_4$ (68%) (4:1)	20°C (12 h) + 100°C (10 min)	Blue mussels	77
HNO_3 (65%) $HClO_4$ (68%) (4:1)	20°C (12 h) + 100°C (10 min)	Brown shrimp	73
CH_2O_2 (3%)	72 h	Corals	99
KOH (10%)	2–3 weeks	Fish	54
KOH (10%)	60°C (12 h)	Fish	21
KOH (10%)	2–3 weeks	Fish	33
H_2O_2 (30%)	60°C	Blue mussels	92
H_2O_2 (30%)	20°C (7 d)	Biogenic matter	100
H_2O_2 (15%)	55°C (3 d)	Fish	72
H_2O_2 (30%)	65°C (24 h) + 20°C (<48 h)	Bivalves	98
NaClO (3%)	20°C (12 h)	Fish	101
$NaClO_3$ (10:1)	20°C (5 min)		
Proteinase K	50°C (2 h)	Zooplankton copepods	9

Assumptions: "Overnight" Given as 12 h; "Room Temperature" Given as 20°C.

7.3.1 Nitric Acid

Nitric acid (HNO_3) is a strong oxidizing mineral acid, capable of molecular cleavage and rapid dissolution of biogenic material. When tested against hydrochloric acid (HCl), hydrogen peroxide (H_2O_2) and sodium hydroxide (NaOH), HNO_3 resulted in the highest digestion efficacies, with >98% weight loss of biological tissue [78]. The optimized protocol involved digesting excised mussel tissue in 69% HNO_3 at room temperature overnight, followed by 2 h at 100°C. Desforges et al. [11] also tested HNO_3, HCl and H_2O_2 in digesting zooplankton and similarly identified nitric acid as the most effective digestion agent based on visual observations; here, the optimized digestion protocol consisted of exposing individual euphausiids to 100% HNO_3 at 80°C for 30 minutes. Adaptations of nitric acid protocols have been successfully used to isolate fibers, films and fragments from a range of organisms [38–39,73–76]. While largely efficacious in digesting organic material, a number of studies observed that oily residue and/or tissue remnants remained postdigestion [18,51,63] which have the potential to obscure microplastics. Consequently, De Witte et al. [77] proposed using a mixture of 65% HNO_3 and 68% perchloric acid ($HClO_4$) in a 4:1 v/v ratio (500 mL acid to 100 g tissue) to digest mussel tissue overnight at room temperature followed by 10 minutes boiling, resulting in the removal of the oily residue. Recovery rates for 10 and 30 mm PS microspheres spiked into mussel tissue and subsequently digested with a nitric acid range between 93.6% and 97.9% [78]. However, the high concentrations of acid and temperatures applied resulted in the destruction of 30–200 mm nylon fibers and melding of 10 mm polystyrene microbeads following direct exposure. Researchers have found that polymeric particles, including polyethylene (PE) and polystyrene (PS), dissolved following overnight exposure and 30 minutes boiling with 22.5 M HNO_3 [72,97]. Polyamide (PA, nylon), polyester (PET) and polycarbonate have low resistance to acids, even at low concentrations. High concentrations of nitric, hydrofluoric, perchloric and sulfuric acid are likely to destroy or severely damage the majority of polymers tested, particularly at higher temperatures. The absence of synthetic fibers in biota digested using HNO_3 is likely a reflection of the destructive power of the acid [98].

7.3.2 Other Acids

Formic and hydrochloric acids (HCl) have also been suggested as digestive agents. With scleractinian corals (*Dipsastrea pallida*), formic acid (3%, 72 h) has been used to decalcify polyps to assist in the visualization of ingested blue polypropylene shavings [99]. HCl has also be tried out as a digestant of microplastics from pelagic and sediment samples; however this non-oxidizing acid proved inconsistent and inefficient in digesting organic material [9].

7.3.3 Alkalis

Excised fish tissues, including the esophagus, stomach and intestines, have been successfully digested using potassium hydroxide (KOH, 10%) following a 2–3 week incubation [33,54]. The protocol has been adapted for the dissolution of gastrointestinal tracts of fish and mussel, crab and oyster tissues, either directly or following baking (450°C, 6 h), by incubating tissues in 10% KOH at 60°C overnight [21,97]. This latter method has proved to be largely efficacious in removing biogenic material. It is well suited to the dissolution of invertebrates and fish fillets, but it is less applicable for fish digestive tracts owing to the presence of inorganic materials. As with HNO_3, an oily residue and bone fragments may remain following digestion. Another strong base, sodium hydroxide (NaOH; 1 M and 10 M), has been successfully applied to remove biogenic material (e.g., zooplankton) from surface trawls, with 90% efficiency based on sample weight loss [9]. Foekema et al.[54] suggests polymers are resistant to KOH, and Dehaut et al. [97] showed no demonstrable impact on polymer mass or form, except in the case of cellulose acetate (CA). Testing the rapid KOH digestion protocol achieved a 100% microplastic recovery rate [97]. PA, PE and polypropylene (PP) are resistant to 10% KOH, but polycarbonate (PC) and PET are degraded. Cole et al. [9] tested 40% NaOH (60°C) on a range of polymers and observed deformation of PA fibers, yellowing of polyvinyl chloride (PVC) granules and melding of polyethylene particles. Dehaut et al. [97] also noted that PC, CA and PET were degraded using this protocol. PE and PVC are resistant to NaOH, even at concentrations of 50% at 50°C.

7.3.4 Oxidizing Agents

Mathalon and Hill [92] used H_2O_2 (30%) at 55°C–65°C to digest mussel soft tissue, and although largely effective, the authors noted "flakes of debris" remained. Li et al. [98] also applied this method, but incubated samples in an oscillating incubator and then at room temperature for 24–48 h. Avio et al. [72] similarly tested alternate treatments in digesting intestinal tracts of mullet (*Mugil cephalus*). Avio et al. [72] identified that direct application of H_2O_2 resulted in only a 70% retrieval of spiked microplastics, with losses linked to H_2O_2 foaming. A number of studies noted excessive foaming might obscure samples or lead to sample loss [72,78,92]. A density separation of stomach contents with hypersaline (NaCl) solution followed by digestion of isolated material with 15% H_2O_2 resulted in a much improved 95% recovery rate for spiked microplastics. Dehaut et al. [97] trialed 0.27 M $K_2S_2O_8$ with 0.24 M NaOH in digesting biological tissues. This mixture was efficient in digesting biogenic material (<99.7% mass reduction), but the authors noted its expense and highlighted issues with crystallization of the digestive solutions and incomplete digestion causing blockages during filtration. Avio et al. [72] observed that 15% H_2O_2 had no visible impact on PE or PS microspheres, although a slight modification to FTIR spectra was observed. Conversely, Nuelle et al. [100] identified some visual deformities to exposed plastic and quantified a 6.2% loss in size for PP and PE particles (<1 mm). $K_2S_2O_8$ resulted in no changes in the mass or appearance in the majority of exposed polymers, but caused complete dissolution of cellulose acetate [97].

Naji et al. [109] applied the digestion method using 30% H_2O_2 to remove soft tissue of different mollusks. This was followed by filtration of the solution through 25 μm filter paper using a vacuum system.

7.3.5 Sodium Hypochlorite

Collard et al. [101] digested fish stomach contents, in an overnight exposure with ~3% NaClO. Filtered digestants were subsequently washed with 65% HNO_3 and digested in a 10:1 NaClO:HNO_3 solution for 5 minutes. The technique caused no visible degradation of a range of polymers (PET, PVC, PE, PP, PS, PC or PA).

7.3.6 Enzymes

To remove biological material from field-collected samples, Cole et al. [9] developed a digestion protocol employing a serine protease (Proteinase K). Material was desiccated (60°C, 24 h), ground and homogenized by repeatedly drawing samples through a syringe. Next, samples were mixed with homogenizing solution (400 mM Tris–HCl buffer, 60 mM EDTA, 105 mM NaCl, 1% SDS), acclimated to 50°C, enzymatically digested with Proteinase K (500 mg/mL per 0.2 g DW sample) and mixed with sodium perchlorate ($NaClO_4$, 5 M). Ultrasonication was demonstrated to have a deleterious effect on digestion efficiency, owing to protein precipitation in the media. With marine samples, the Proteinase K method proved to have a digestion efficacy of >97%, and the method was used to isolate fluorescent polystyrene microspheres (20 mm) ingested by marine copepods. The authors note that additional enzymes could be used depending on the chemical makeup of the organism or samples in question (e.g., chitinase with chitinous invertebrates). The enzyme pepsin causes no damage to polymers, but proved only partially effective at digesting biogenic material [100]. Enzymes have been successfully applied recently in the isolation of microplastics from the intestinal tracts of turtles with Proteinase K [102]; mussel tissue with Corolase 7089 (AB enzymes) [103] and herring digestive tracts with Proteinase K and H_2O_2 [104]. In contrast to chemical digestion techniques, enzymes ensure no loss, degradation or surface change to plastics present [9] and are less hazardous to human health.

Table 7.2 summarizes the effects of enzyme digestion on *Mytilus trossulus* [108].

7.3.7 Effect of Temperature

The performance of digesting solutions across different temperatures was systematically investigated by Karami et al [106]. In the first phase, the efficiency of different oxidative agents (NaClO or H_2O_2), bases (NaOH or KOH) and acids (HCl or HNO_3; concentrated and diluted [5%]) in digesting fish tissues at room temperature (RT, 25°C), 40°C, 50°C or 60 °C were measured. In the second phase, the treatments that were efficient in digesting the biological materials were evaluated for their compatibility with eight major plastic polymers (assessed through recovery rate method for the water samples). The isolation of microplastics from Asian clams was based on previous studies [98,107].

7.4 FILTERING DIGESTANTS

Following digestion, chemical agents can be filtered to retain any recalcitrant materials (e.g., undigested tissue, inorganic residue, microplastics). Viable filters include 0.2 and 0.7 mm glass fiber filters [38,74], 5 mm cellulose nitrate filters [75], 5 mm cellulose acetate membranes [101], 50 mm mesh [9] and 250 mm mesh [31]. Larger pore size facilitates rapid filtering but will result in the loss of smaller plastics [9]. Glass fiber filters can shed and might be considered a source of contamination; smoother filters (e.g., membrane filters) are typically easier to scrape and less prone to fragment. Microplastics on filters can be visualized directly, transferred to slides or extracted. Collard et al. [101] suggests placing filters in methanol solution, ultrasonicating (50 Hz), centrifuging (5000 rpm, 5 min, 20°C) and then removing pelleted plastic by pipette; while this method was suitable for a range of polymers, the methanol caused a 25% weight reduction in tested PVC particles.

TABLE 7.2   Comparison of Tested Mussel Tissue Digestion Methods

Test Number	Solution	Temp (°C)	Time (hrs)	Results	Notes
1	NaOH (20 g/L)	60	24	• Mussel tissue dissolved into yellow liquid • Some 2–3 mm particles remain • Nonsynthetic fibers start to leach color in an hour	• Hazardous • Damages glassware • Visible digestion effects in an hour; tissue surface is disintegrating • Samples pass through glass microfiber filter effortlessly
2	NaOH (20 g/L)+SDS (5 g/L)	60	24	Dissolved tissues • Foggy solution • Nonsynthetic fibers start to leach color in an hour	• See #1
3	NaOH (20 g/L) + NaClO (14%)	60	24	• See #1	• See #1
4	NaClO (14%) + SDS (5 g/L)	60	24	• Tissue digested nearly completely	• See #1
5	KOH (10%)	50	24	• Dissolved into liquid with several 2 mm pieces of tissue	• Hazardous • Clogs glass fiber filters
6	KOH (10%) + SDS (5 g/L)	50	24	• See #5	• Sample turned into jelly when left in room temperature for 30 min after oven incubation • SDS seems to aid the digestion process slightly • See #5
7	Enzyme mix (1:1 Biozym F: Biozym SE)	37.5	48–72	• Deep yellow, dense solution with a few organic particles remaining • Fibers not affected	• Works well with 20 μm plankton net filter, clogs glass fiber filter • User friendly (not hazardous to use) • Occasional remaining tissue particles break upon filtration
8	SDS (5 g/L) + Enzymes mix (1:1 Biozym F: Biozym SE) 24 h later	SDS: 70 Enzymes: lowered to 37.5	24 (SDS) + 48 (enzym.)	• See #7	• Contamination risk increased due to additional handling of samples • Risk of enzyme denaturation if solution not cool enough upon adding the enzymes • No clear difference in digestion speeds between protocol #8 and #9 • See #7
9	1:1 SDS (5 g/L): Enzymes mix (1:1 Biozym F: Biozym SE)	37.5	48–72	• See #7	• Addition of SDS partially prevents particles from attaching to vial walls • See #7

Source: Railo S et al. *Mar Pollut Bull* 2018, 130, 206–214.

7.5 DENSITY SEPARATION

Density separation was performed by NaCl [72,92,98] or by centrifuge [101]. Following settlement of denser materials, the supernatant is filtered, and the resulting material examined under microscope. Density separation can be useful in studies following digestion. Saturated salt solutions, such as NaCl (aq) allow the separation of less dense particles where there are large amounts of inorganic matter (e.g., sand, chitin, bone) that have not been dissolved [1]. Density separation has been recommended by the MSFD (EU) for Europe. NaCl is recommended because it is inexpensive and nonhazardous; however, the use of NaCl could lead to an underestimation of more-dense particles (>1.2 g/cm^3). NaI and ZnCl$_2$ solutions have been considered as viable alternatives to NaCl (aq) [105]. Their high density makes them capable of floating high-density plastics including PVC.

REFERENCES

1. Lusher AL, Welden NA, Sobral P, and Cole M. Sampling, isolating and identifying microplastics ingested by fish and invertebrates. *Anal Methods* 2017, 9, 1346–1360.
2. Lusher A. In: Bergmann M, Gutow L, and Klages M (eds) *Marine Anthropogenic Litter.* Springer, Berlin, 2015, ch. 10, pp. 245–308.
3. Plastics – the Facts 2014/2015, http://issuu.com/plasticseuropeebook/docs/final_plastics_the_facts_2014_19122, accessed May 2019.
4. Cole M, Lindeque P, Halsband C, Galloway. TS, Microplastics as contaminants in the marine environment: a review. *Mar Pollut Bull.* 2011, 62(12), 2588–2597.
5. UNEP, Marine Plastic Debris and Microplastic Technical Report, United Nations Environmental Programme, Nairobi, 2016.
6. Murray F and Cowie PR. Plastic contamination in the decapod crustacean *Nephrops norvegicus* (Linnaeus, 1758). *Mar Pollut Bull* 2011, 62(6), 1207–1217.
7. Welden NA and Cowie PR. Environment and gut morphology influence microplastic retention in langoustine, *Nephrops norvegicus. Environ Pollut* 2016, 214, 859–865.
8. Cole M, Lindeque P, Fileman E, Halsband C, Goodhead R, Moger J, and Galloway TS. Microplastic Ingestion by Zooplankton. *Environ Sci Technol* 2013, 47(12), 6646–6655.
9. Cole M, Webb H, Lindeque PK, Fileman ES, Halsband C, and Galloway TS. Isolation of microplastics in biota-rich seawater samples and marine organisms. *Sci Rep* 2014, 4, 4528,
10. Cole M, Lindeque H, Fileman ES, Halsband C, and Galloway TS. The Impact of Polystyrene Microplastics on Feeding, Function and Fecundity in the Marine Copepod Calanus helgolandicus. *Environ Sci Technol.* 2015, 49(2), 1130–1137.
11. Desforges JPW, Galbraith M, and Ross PS. Ingestion of Microplastics by Zooplankton in the Northeast Pacific Ocean. *Arch Environ Contam Toxicol* 2015, 69(3), 320–330.
12. Faure F, Demars C, Wieser O, Kunz M, and De Alencastro LF. Plastic pollution in Swiss surface waters: nature and concentrations, interaction with pollutants. *Environ Chem* 2015, 12(5), 582–591.
13. Browne MA, Niven SJ, Galloway TS, Rowland SJ, and Thompson RC. Microplastic Moves Pollutants and Additives to Worms, Reducing Functions Linked to Health and Biodiversity. *Curr Biol* 2013, 23(23), 2388–2392.
14. Watts AJ, Lewis C, Goodhead RM, Beckett SJ, Moger J, Tyler CR, and Galloway TS. Uptake and Retention of Microplastics by the shore crab Carcinus maenas. *Environ Sci Technol* 2014, 48(15), 8823–8830.

15. Watts AJ, Urbina MA, Corr S, Lewis C, and Galloway TS. Ingestion of Plastic Microfibers by the Crab *Carcinus maenas* and Its Effect on Food Consumption and Energy Balance. *Environ Sci Technol* 2015, 49(24), 14597–14604.

16. Hämer J, Gutow L, Köhler A, and Saborowski R. Fate of Microplastics in the Marine Isopod Idotea emarginata. *Environ.Sci Technol* 2014, 48(22), 13451–13458.

17. Canesi L, Ciacci C, Bergami E, Monopoli MP, Dawson KA, Papa S, Canonico B, and Corsi I. Evidence for immunomodulation and apoptotic processes induced by cationic polystyrene nanoparticles in the hemocytes of the marine bivalve *Mytilus*. *Mar Environ Res* 2015, 111, 34–40

18. Cole M and Galloway TS. Ingestion of Nanoplastics and Microplastics by Pacific Oyster Larvae. *Environ Sci Technol* 2015, 49(24), 14625–14632.

19. Avio CG, Gorbi S, Milan M, Benedetti M, Fattorini D, d'Errico G, Pauletto M, Bargelloni L, and Regoli F, Pollutants bioavailability and toxicological risk from microplastics to marine mussels. *Environ Pollut* 2015, 198, 211–222.

20. Canesi L, Ciacci C, Bergami E, Monopoli MP, Dawson KA Papa S, Canonico B, and Corsi I. Evidence for immunomodulation and apoptotic processes induced by cationic polystyrene nanoparticles in the hemocytes of the marine bivalve *Mytilus*. *Mar Environ Res* 2015, 111, 34–40.

21. Rochman CM, Tahir A, Williams SL, Baxa DV, Lam R, Miller JT, Teh FC, Werorilangi S, and Teh SJ. Anthropogenic debris in seafood: Plastic debris and fibers from textiles in fish and bivalves sold for human consumption. *Sci Rep* 2015, 5, 14340

22. Biginagwa FJ, Mayoma BS, Shashoua Y, Syberg K, and Khan FR. First evidence of microplastics in the African Great Lakes: Recovery from Lake Victoria Nile perch and Nile tilapia. *J Great Lakes Res* 2016, 42(1), 146–149.

23. Dos Santos J and Jobling M. A model to describe gastric evacuation in cod (*Gadus morhua* L.) fed natural prey. *ICES J Mar Sci* 1992, 49(2), 145–154.

24. Sæther BS and Jobling M. Gastrointestinal evacuation of inert particles by turbot, Psetta maxima: evaluation of the X-radiographic method for use in feed intake studies. *Aquat Living Resour* 1997, 10(6), 359–364.

25. McGaw IJ and Curtis DL. A review of gastric processing in decapod crustaceans. *J Comp Physiol., B* 2013, 183(4), 443–465.

26. Powell MD and Berry AJ. Ingestion and regurgitation of living and inert materials by the estuarine copepod *Eurytemora affinis* (Poppe) and the influence of salinity. Estuarine, *Coastal Shelf Sci* 1990, 31(6), 763–773.

27. Bromley PJ, The role of gastric evacuation experiments in quantifying the feeding rates of predatory fish. *Rev Fish Biol Fish* 1994, 4(1), 36–66.

28. Bowman RE. Effect of regurgitation on stomach content data of marine fishes. *Environ Biol Fishes*, 1986, 16(1), 171–181.

29. Bowen SH. In: Murphy BR and Willis DW (eds) *Fisheries Techniques*. American Fisheries Society, Bethesda, MD, 2nd edn, 1996, pp. 513–532.

30. Daan N. A quantitative analysis of the food intake of North Sea cod, Gadus Morhua. *Neth J Sea Res* 1973, 6(4), 479–517.

31. Lusher AL, McHugh M, and Thompson RC. Occurrence of microplastics in the gastrointestinal tract of pelagic and demersal fish from the English Channel. *Mar Pollut Bull* 2013, 67(1–2), 94–99.

32. Neves D, Sobral P, Ferreira JL, and Pereira T. Ingestion of microplastics by commercial fish off the Portuguese coast. *Mar Pollut Bull* 2015, 101, 119–126.

33. Lusher AL, O'Donnell C, Officer R, and O'Connor I, Microplastic interactions with North Atlantic mesopelagic fish. *ICES J Mar Sci* 2016, 73(4), 1214–1225.

34. Cole M, Lindeque P, Fileman ES, Halsband C, Galloway TS. Microplastics Alter the Properties and Sinking Rates of Zooplankton Faecal Pellets. *Environ Sci Technol* 2016, 50(6), 3239–3246.

35. Kaposi KL, Mos B, Kelaher BP, and Dworjanyn SA. Ingestion of Microplastic Has Limited Impact on a Marine Larva. *Environ Sci Technol* 2014, 48(3), 1638–1645.

36. Browne MA, Dissanayake A, Galloway TS, Lowe DM, Thompson RC. Ingested Microscopic Plastic Translocates to the Circulatory System of the Mussel, Mytilus edulis (L.). *Environ Sci Technol* 2008, 42(13), 5026–5031.

37. Von Moos N, Burkhardt-Holm P and Köhler A. Uptake and Effects of Microplastics on Cells and Tissue of the Blue Mussel Mytilus edulis L. after an Experimental Exposure. *Environ Sci Technol* 2012, 46(20), 11327–11335.

38. Lu Y, Zhang Y, Deng Y, Jiang W, Zhao Y, Geng J, Ding L, and Ren H. Uptake and Accumulation of Polystyrene Microplastics in Zebrafish (Danio rerio) and Toxic Effects in Liver. *Environ Sci Technol* 2016, 50(7), 4054–4060.

39. Braid HE, Deeds J, DeGreasse SL, Wilson JJ, Osborne J and Hanner RH. Preying on commercial fisheries and accumulating paralytic shellfish toxins: a dietary analysis of invasive *Dosidicus gigas* (Cephalopoda Ommastrephidae) stranded in Pacific Canada. *Mar Biol* 2012, 159(1), 25–31.

40. Besseling E, Foekema EM, van Franeker JA, Leopold MF, Kühn S, Rebolledo EB, Heße E, Mielke L, IJzer J, Kamminga P, and Koelmans AA. Microplastic in a macro filter feeder: Humpback whale *Megaptera novaeangliae*. *Mar Pollut Bull* 2015, 95(1), 248–252.

41. Lusher AL, Hernandez-Milian G, O'Brien J, Berrow S, O'Connor I, and Officer R. Microplastic and macroplastic ingestion by a deep diving, oceanic cetacean: The True's beaked whale Mesoplodon mirus. *Environ Pollut* 2015, 199, 185–191.

42. Tourinho PS, do Sul JAI and Fillmann G. Is marine debris ingestion still a problem for the coastal marine biota of southern Brazil? *Mar Pollut Bull* 2010, 60(3), 396–401.

43. Van Franeker JA, Blaize C, Danielsen J, Fairclough K, Gollan J, Guse N, Hansen PL, Heubeck M, Jensen JK, Le Guillou G and Olsen B. Monitoring plastic ingestion by the northern fulmar *Fulmarus glacialis* in the North Sea. *Environ Pollut* 2011, 159(10), 2609–2615.

44. Thompson RC, Olsen Y, Mitchell RP, Davis A, Rowland SJ, John AW, McGonigle D and Russell AE. Lost at Sea: Where Is All the Plastic? *Science*. 2004, 304(5672), 838.

45. Goldstein MC and Goodwin DS. Gooseneck barnacles (Lepas spp.) ingest microplastic debris in the North Pacific Subtropical Gyre. *Peer J* 2014, 1, e841.

46. Carpenter EJ, Anderson SJ, Harvey GR, Miklas HP, and Peck BB. Polystyrene spherules in coastal waters. *Science*, 1972, 178, 749–750.

47. Kartar S, Milne RA and Sainsbury M. Polystyrene waste in the Severn Estuary. *Mar Pollut Bull*, 4(9), 144.

48. Boerger CM, Lattin GL, Moore SL and Moore CJ. Plastic ingestion by planktivorous fishes in the North Pacific Central Gyre. *Mar Pollut Bull* 2010, 60(12), 2275–2278.

49. Davison P and Asch RG. Plastic ingestion by mesopelagic fishes in the North Pacific Subtropical Gyre. *Mar Ecol: Prog Ser* 2011, 432, 173–180.

50. Possatto FE, Barletta M, Costa MF, do Sul JAI and Dantas DA. Plastic debris ingestion by marine catfish: An unexpected fisheries impact. *Mar Pollut Bull* 2011, 62(5), 1098–1102.

51. Dantas DV, Barletta M and Costa MF. The seasonal and spatial patterns of ingestion of polyfilament nylon fragments by estuarine drums (Sciaenidae). *Environ Sci Pollut Res* 2012, 19(2), 600–606.

52. Ramos JAA, Barletta M and Costa MF. Ingestion of nylon threads by Gerreidae while using a tropical estuary as foraging grounds. *Aquat Biol* 2012, 17, 29–34.

53. Choy CA and Drazen JC. Plastic for dinner? Observations of frequent debris ingestion by pelagic predatory fishes from the central North Pacific. *Mar Ecol.: Prog Ser* 2013, 485, 155–163.

54. Foekema EM, De Gruijter C, Mergia MT, van Franeker JA, Murk AJ and Koelmans AA. Plastic in North Sea Fish Environ. *Sci Technol* 2013, 47(15), 8818–8824.

55. Gassel M, Harwani S, Park J-S and Jahn A. Detection of nonylphenol and persistent organic pollutants in fish from the North Pacific Central Gyre. *Mar Pollut Bull* 2013, 73, 231–242.

56. Lusher AL, McHugh M and Thompson RC. Occurrence of microplastics in the gastrointestinal tract of pelagic and demersal fish from the English Channel. *Mar Pollut Bull* 2013, 67(1–2), 94–99.

57. Saji Kumar KK, Ragesh N, Remya R and Mohamed KS. Occurrence of plastic debris in the stomach of yellowfin tuna (Thunnus albacares) from the Arabian Sea: A cause for concern. *Marine Fisheries Information Service; Technical and Extension Series* 2013, 217, 13.

58. Kripa V, Nair PG, Dhanya AM, Pravita VP, Abhilash KS, Mohamed AA, Vijayan D, Vishnu PG, Morhan G, Anilkkumar PS and Khambadkar LR. *Microplastics in the gut of anchovies caught from the mud bank area of Alappuzha, Kerala. Marine Fisheries Information Service; Technical and Extension Series* 2014, 219, 27–128.

59. Sulochanan B, Bhat GS, Lavanya S, Dineshbabu AP and Kaladharan P. A preliminary assessment of ecosystem process and marine litter in the beaches of Mangalore. *Indian J Geo-Mar Sci* 2013, 43(9), 1–6.

60. Phillips MB and Bonner TH. Occurrence and amount of microplastic ingested by fishes in watersheds of the Gulf of Mexico. *Mar Pollut Bull* 2015, 100, 264–269.

61. Romeo T, Pietro B, Peda C, Consoli P, Andaloro F, Fossi MC. First evidence of presence of plastic debris in stomach of large pelagic fish in the Mediterranean Sea. *Mar Pollut Bull* 2015, 95, 358–361.

62. Bellas J, Martínez-Armental J, Martínez-Cámara A, Besada V and Martínez-Gmóez C. Ingestion of microplastics by demersal fish from the Spanish Atlantic and Mediterranean coasts. *Mar Pollut Bull* 2016, 109(1), 55–60.

63. Cannon SME, Lavers JL and Figueiredo B. Plastic ingestion by fish in the Southern Hemisphere: A baseline study and review of methods. *Mar Pollut Bull* 2016, 107, 286–291.

64. Miranda DDA and de Carvalho-Souza GF. Are we eating plastic-ingesting fish? *Mar Pollut Bull* 2016, 103, 109–114.

65. Lusher AL, O'Donnell C, Officer R and O'Connor I. Microplastic interactions with North Atlantic mesopelagic fish. *ICES J Mar Sci* 2016, 73(4), 1214–1225.

66. Nadal MA, Alomar C and Deudero S. High levels of microplastic ingestion by the semipelagic fish bogue *Boops boops* (L.) around the Balearic Islands. *Environ Pollut* 2016, 214, 517–523.

67. Naidoo T, Smit AJ and Glassom D. Plastic pollution in five urban estuaries of KwaZulu-Natal, South Africa. *Afr J Mar Sci* 2016, 38(1), 145–149.

68. Pedà C, Caccamo L, Fossi MC, Gai F, Andaloro F, Genovese L, Perdichizzi A, Romeo T and Maricchiolo G. Intestinal alterations in European sea bass *Dicentrarchus labrax* (Linnaeus, 1758) exposed to microplastics: Preliminary results. *Environ Pollut* 2016, 212, 251–256.

69. Barltrop D and Meek F. Effect of Particle Size on Lead Absorption from the Gut. *Arch Environ Health*, 1979, 34(4), 280–385.

70. Reineke JJ, Cho DY, Dingle Y-T, Morello AP, Jacob J, Thanos CG and Mathiowitz E. Unique insights into the intestinal absorption, transit, and subsequent biodistribution of polymer-derived microspheres. *Proc Natl Acad Sci U S A* 2013, 110(34), 13803–13808.

71. Paul-Pont I, Lacroix C, Gonzalez Fernandez C, Hegaret H, Lambert C, Le Goic N, Frere L, Cassone A-L, Sussarellu R, Fabioux C, Guyomarch J, Albentosa M, Huvet A and Soudant P. Exposure of marine mussels *Mytilus* spp. to polystyrene microplastics: Toxicity and influence on fluoranthene bioaccumulation. *Environ Pollut* 2016, 216, 724–737.

72. Avio CG, Gorbi S and Regoli F. Experimental development of a new protocol for extraction and characterization of microplastics in fish tissues: First observations in commercial species from Adriatic Sea. *Mar Environ Res* 2015, 111, 18–26.

73. Devriese LI, van der Meulen MD, Maes T, Bekaert K, Paul-Pont I, Frère L, Robbens J and Vethaak AD. Microplastic contamination in brown shrimp (*Crangon crangon*, Linnaeus 1758) from coastal waters of the Southern North Sea and Channel area. *Mar Pollut Bull* 2015, 98(1–2), 179–187.

74. Davidson K and Dudas SE. Microplastic ingestion by wild and cultured Manila Clams (*Venerupis philippinarum*) from Baynes Sound, British Columbia. *Arch Environ Contam Toxicol* 2016, 71(2), 147–156.

75. Van Cauwenberghe L and Janssen CR. Microplastics in bivalves cultured for human consumption. *Environ Pollut* 2014, 193, 65–70.

76. Van Cauwenberghe L, Claessens M, Vandegehuchte MB and Janssen CR. Microplastics are taken up by mussels (*Mytilus edulis*) and lugworms (*Arenicola marina*) living in natural habitats. *Environ Pollut* 2015, 199, 10–17.

77. De Witte B, Devriese L, Bekaert K, Hoffman S, Vandermeersch G, Cooreman K and Robbens J. Quality assessment of the blue mussel (*Mytilus edulis*): Comparison between commercial and wild types. *Mar Pollut Bull* 2014, 85(1), 146–155.

78. Claessens M, Van Cauwenberghe L, Vandegehuchte MB and Janssen CR. New techniques for the detection of microplastics in sediments and field collected organisms. *Mar Pollut Bull* 2013, 70(1–2), 227–233.

79. Graham ER and Thompson JT. Deposit- and suspension-feeding sea cucumbers (Echinodermata) ingest plastic fragments. *J Exp Mar Biol Ecol* 2009, 368(1), 22–29.

80. Setälä O, Flemming-Lehtinen V and Lehtiniemi M., Ingestion and transfer of microplastics in the planktonic food web Environ. *Pollut* 2014, 185, 77–83

81. Ugolini A, Ungherese G, Ciofni M, Lapucci A and Camaiti M. Microplastic debris in sandhoppers. Estuarine, *Coastal Shelf Sci* 2013, 129, 19–22.

82. Ward JE and Targett NM. Influence of marine microalgal metabolites on the feeding behavior of the blue mussel *Mytilus edulis*. *Mar Biol* 1989, 101(3), 313–321.

83. Brilliant MGS and MacDonald BA. Postingestive selection in the sea scallop, *Placopecten magellanicus* (Gmelin): the role of particle size and density. *J Exp Mar Biol Ecol* 2000, 253(2), 211–227.

84. Brillant M and MacDonald B. Postingestive selection in the sea scallop (*Placopecten magellanicus*) on the basis of chemical properties of particles. *Mar Biol* 2002, 141(3), 457–465.

85. Ward JE, Levinton JS and Shumway SE. Particle sorting in bivalves: *in vivo* determination of the pallial organs of selection. *J Exp Mar Biol Ecol* 2003, 293(2), 129–149.

86. Browne MA, Dissanayake A, Galloway TS, Lowe DM and Thompson RC. Ingested Microscopic Plastic Translocates to the Circulatory System of the Mussel, *Mytilus edulis* (L.). *Environ Sci Technol* 2008, 42(13), 5026–5031.

87. Ward JE and Kach DJ. Marine aggregates facilitate ingestion of nanoparticles by suspension-feeding bivalves. *Mar Environ Res* 2009, 68(3), 137–142.

88. Von Moos N, Burkhardt-Holm P and Köhler A. Uptake and Effects of Microplastics on Cells and Tissue of the Blue Mussel *Mytilus edulis* L. after an Experimental Exposure. *Environ Sci Technol* 2012, 46(20), 11327–11335.

89. Wegner A, Besseling E, Foekema EM, Kamermans P and Koelmans AA. Effects of nanopolystyrene on the feeding behavior of the blue mussel (*Mytilus edulis* L.). *Environ Toxicol Chem* 2012, 31, 2490–2497.

90. Claessens M, Van Cauwenberghe L, Vandegehuchte MB and Janssen CR. New techniques for the detection of microplastics in sediments and field collected organisms. *Mar Pollut Bull* 2013, 70(1–2), 227–233.

91. De Witte B, Devriese L, Bekaert K, Hoffman S, Vandermeersch G, Cooreman K and Robbens J. Quality assessment of the blue mussel (*Mytilus edulis*): Comparison between commercial and wild types. *Mar Pollut Bull* 2014, 85(1), 146–155.

92. Mathalon A and Hill P. Microplastic fibers in the intertidal ecosystem surrounding Halifax Harbor, Nova Scotia. *Mar Pollut Bull* 2014, 81(1), 69–79.

93. Van Cauwenberghe L and Janssen CR, Microplastics in bivalves cultured for human consumption. *Environ Pollut* 2014, 193, 65–70.

94. Avio CG, Gorbi S, Milan M, Benedetti M, Fattorini D, d'Errico G, Pauletto M, Bargelloni L and Regoli F. Pollutants bioavailability and toxicological risk from microplastics to marine mussels. *Environ Pollut* 2015, 198, 211–222.

95. Wright SL, Rowe D, Thompson RC and Galloway TS. Microplastic ingestion decreases energy reserves in marine worms. *Curr Biol* 2013, 23(23), R1031–R1033.

96. Andrady AL. In: Bergmann M, Gutow L and Klages M (eds) *Marine Anthropogenic Litter*. Springer, Berlin, 2015, ch. 3, pp. 57–72.

97. Dehaut A, Cassone A-L, Frere L, Hermabessiere L, Himber C, Rinnert E, Riviere G, Lambert C, Soudant P, Huvet A, Duflos G and Paul-Pont I. Microplastics in seafood: Benchmark protocol for their extraction and characterization. *Environ Pollut* 2016, 215, 223–233.

98. Li J, Yang D, Li L, Jabeen K and Shi H. Microplastics in commercial bivalves from China. *Environ Pollut* 2015, 207, 190–195.

99. Hall NM, Berry KL E, Rintoul L and Hoogenboom MO. Microplastic ingestion by scleractinian corals. *Mar Biol* 2015, 162(3), 725–732.

100. Nuelle MT, Dekiff JH, Remy D and Fries E. A new analytical approach for monitoring microplastics in marine sediments. *Environ Pollut* 2014, 184, 161–169.

101. Collard F, Gilbert B, Eppe G, Parmentier E and Das K. Detection of anthropogenic particles fish stomachs: An isolation method adapted to identification by Raman Spectroscopy Arch. *Environ Contam Toxicol* 2015, 69(3), 331–339.

102. Duncan E. Investigating the Presence and Effects of Microplastics in Sea Turtles presented in part at Micro 2016, Lanzarote, Canary Islands, Spain, May, 2016.

103. Catarino AI, Thompson R, Sanderson W and Henry TB. Development and optimization of a standard method for extraction of microplastics in mussels by enzyme digestion of soft tissues. *Environ Toxicol Chem* 2017, 36 (4), 947–951.

104. Santana MFM, Ascer LG, Custodio MR, Moreira FT and Turra A. Microplastic contamination in natural mussel beds from a Brazilian urbanized coastal region: Rapid evaluation through bioassessment. *Mar Pollut Bull* 2016, 106(1–2), 183–189.

105. Horton AA, Svendsen C, Williams RJ, Spurgeon DJ and Lahive E. Large microplastic particles in sediments of tributaries of the River Thames, UK – Abundance, sources and methods for effective quantification. *Mar Pollut Bull* 2016, 114 (1), 218–226.
106. Karami A, Golieskardi A, Choo CK, Romani N, Ho YB, Salamatinia B. A high-performance protocol for extraction of microplastics in fish. *Sci Total Environ* 2017, 578, 485–494.
107. Su L, Xue Y,. Li L, Yang D, Kolandhasamy P, Li D. Microplastics in Taihu Lake, China. *Environ Pollut* 2016, 216, 711–719.
108. Railo S, Talvitie J, Setälä O, Koistinen A, Lehttiniemi M. Application of an enzyme digestion method reveals microlitter in Mytilus trossulus at a wastewater discharge area. *Mar Pollut Bull* 2018, 130, 206–214.
109. Naji A, Nuri M, and Vethaak AD. Microplastics contamination in molluscs from the northern part of the Persian Gulf. *Environ Pollut* 2018, 235, 113–120.

Sampling, Isolating and Identifying Microplastics Ingested by Fish and Invertebrates*

A. L. Lusher, N. A. Welden, P. Sobral and M. Cole

CONTENTS

* Previously published in *Anal. Methods*, 2017, 9, 1346-1360 (DOI: 10.1039/C6AY02415G). This Open Access Article is licensed under a Creative Commons Attribution 3.0 Unported Licence.

8.1 INTRODUCTION

Over the past century, there has been an exponential increase in plastic demand and production [1]. Concurrently, improper disposal, accidental loss and fragmentation of plastic materials have led to an increase in tiny plastic particles and fibers (microplastic, <5 mm) polluting the environment [2,3]. Microplastics have been observed in marine [4], freshwater [5,6] and terrestrial [7] ecosystems across the globe, and biotic interactions are widely evidenced (Figure 8.1). Microplastics can be consumed by a diverse array of marine organisms across trophic levels, including protists [8], zooplankton [9–17] annelids [18–26], echinoderms [27–31], cnidarian [32], amphipods [19,26,33], decapods [34–41], isopods [42], bivalves [43–60], cephalopods [61], barnacles [62], fish [58,66–94] turtles [95], birds [96] and cetaceans [97,98]. Over 220 different species have been found to consume microplastic debris *in natura*. Of these, ingestion is reported in over 80% of the sampled populations of some invertebrate species [34,38,41]. Interactions between microplastics and freshwater invertebrates, fish and birds are increasingly reported [99–107], although some researchers are focusing on model species such as *Daphnia magna* [108–111]. The consumption of microplastics by terrestrial organisms is poorly documented; however, laboratory studies indicate earthworms (*Lumbricus terrestris*) can consume plastic particles present in soil [112].

There are a number of exposure pathways by which organisms may interact with microplastic debris. Direct consumption of microplastic is prevalent in suspension feeders, including zooplankton [10], oysters [59] and mussels [43–58,60–63], and deposit feeders, such as sea cucumbers [28], crabs [35–37,39,40] and *Nephrops*, [34,41] owing to their inability to differentiate between microplastics and prey. Predators and detritivores may indirectly ingest plastic while consuming prey (i.e., trophic transfer) or scavenging detrital matter (e.g., marine snows, fecal pellets, carcasses) containing microplastic [13,34,35,41,113–115]. Micro- and nanoplastics can adhere to external

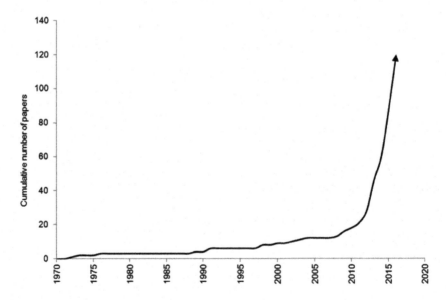

FIGURE 8.1   Publication trend of studies investigating biota interactions with microplastics through June 30, 2016.

appendages, including the gills of the shore crab (*Carcinus maenas*) [37] and mussels (*Mytilus edulis*) [62] and setae of copepod swimming legs and antennules [10]. Other studies have identified that microplastics can bind to microalgae [116–118] or macroalgae [119]. Microplastic exposure has been associated with a suite of negative health effects, including increased immune response [49], decreased food consumption [20,22], weight loss [20], decreased growth rate [112], decreased fecundity [59], energy depletion [22] and negative impacts on subsequent generations [59,104]. Microplastics have also been shown to readily accumulate waterborne persistent organic pollutants including pesticides, solvents and pharmaceuticals, which may pose further health effects such as endocrine disruption and morbidity [106,120,121].

The United Nations Environment Programme (UNEP) has identified plastic pollution as a critical problem; the scale and degree of this environmental issue is comparable to that of climate change [3]. There is currently much public and political debate surrounding the issue of microplastics as additives to household and industrial products and the methods by which impacts of said microplastics on the environment are to be measured. Determining the degree to which biota consume microplastics is essential to determine and monitor "good environmental status" for plastic pollution (e.g., EU Marine Strategy Framework Directive, 2008/56/EC; UNEA, US EPA). Equally, the development of robust environmental legislation is reliant on toxicological studies with ecological relevance, requiring an accurate measure of microplastic loads *in natura* [122]. As such, it is imperative that researchers are able to accurately isolate, identify and enumerate microplastic debris consumed by or entangled with biota. Here, systematically and critically review methods employed in the extraction, identification and quantification of microplastic particles ingested by biota are reviewed. The effectiveness and limitations of a range of field sampling, laboratory exposure, extraction, and analytical techniques, and steps for mitigating contamination, are considered. This review primarily focuses on peer-reviewed publications that have investigated the interactions between invertebrates and fish from the wild and following controlled laboratory exposure.

8.2 METHODOLOGICAL REVIEW

For this literature review, original peer-reviewed research articles, gray literature and conference proceedings from the 1970s to July 2016 were examined. Literature referring to the extraction of microplastics from marine, freshwater and terrestrial biota using Web of Knowledge, Science Direct Scopus and Google Scholar were identified. The authors also mined the journals *Marine Pollution Bulletin*, *Environmental Pollution* and *Environmental Science and Technology*, owing to the regularity with which they publish relevant material. Analysis of microplastic specific studies was expanded to include historical literature that did not necessarily have microplastics as the central theme of the research, such as studies which used fluorescent latex beads as a tracer for feeding and retention experiments. Of the 120 papers included in this meta-analysis, 58.3% of studies were conducted in the laboratory, 38.3% focused on organisms collected from the wild and 3.4% involved both laboratory exposure and field collection (Figure 8.2). There were 96 studies wholly focused on marine organisms, 21 on freshwater, two studies on both marine and freshwater organisms and one published study on a terrestrial species.

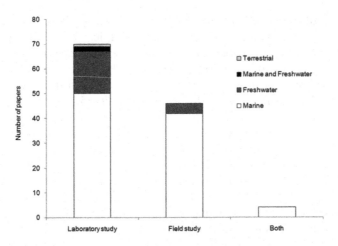

FIGURE 8.2   Studies of biota interactions with microplastic in the laboratory and field.

8.2.1 Sampling

8.2.1.1 Field-Collected Organisms

Observations of microplastic uptake by environmentally exposed organisms have now been reported in a range of habitats, including the sea surface, water column, benthos, estuaries, beaches and aquaculture [4]. The diversity of the organisms studied and the habitats from which they are sampled require a range of collection techniques [123]: the sampling method employed is determined by the research question, available resources, habitat and target organism. Benthic invertebrate species such as *Nephrops norvegicus* may be collected in grabs, traps and creels, or by bottom trawling [34,41], and planktonic and nektonic invertebrates by way of manta and bongo nets [10,12,14,16]. Fish species are generally recovered in surface, midwater and benthic trawls, depending on their habitats [69–92]. Gill nets have been used in riverine systems [102]. Some species are collected from the field by hand; this is common practice for bivalves, crustaceans and annelids [21,35,37,42,56]. Another method is direct collection from shellfish or fish farms [15,55,56] or from commercial fish markets, where the capture method is often unknown [58,103]. Avoiding contamination and biases during sampling and sample analysis is paramount, and mitigation protocols are described below.

8.2.1.1.1 Microplastic Losses during Field Sampling

Handling stress, physical movement and the physiological and behavioral specificities of the sampled organism, may result in the loss of microplastics prior to animal preservation. Gut evacuation times for animals are varied, ranging from as little as 30 minutes for decapod crustaceans (N. Welden, personal observations), <2 hours for calanoid copepods [10], 10–52 hours for fish [68,124] to over 150 hours in larger lobsters [125]. Therefore, some animals might egest microplastic debris prior to analysis [41]. In such cases, the time between sample collection and the preservation of the animal must be as short as possible.

Care must also be taken to minimize handling stress or physical damage. This will reduce the potential for microplastic regurgitation; the frequency with which animals expel consumed plastics during sampling is unknown. The copepod *Eurytemora affinis* [126] and some fish species have been observed regurgitating their stomach contents [127]. The main cause of regurgitation in fish is thought to be related to the expansion of gas in the swim bladder: this causes the compression of the stomach and may, in extreme cases, result in total stomach inversion [128]. Compression of a catch in the cod end might induce regurgitation in fish [129]. The likelihood of regurgitation increases with depth of capture, and gadoids are more prone to regurgitation than flatfish. Piscivorous predators are prone to regurgitation owing to their large distensive esophagus and stomach [128,130]. As such, regurgitation may bias the stomach content estimation, affecting consumption estimates and the presence of plastic debris.

8.2.1.1.2 Microplastic Accumulation during Field Sampling

Laboratory studies have identified that nano- and microplastics can adhere to external appendages of marine copepods [10]. Cataloguing such interactions *in natura* is complicated as determining whether the resulting accumulation has occurred naturally, or as a by-product of the sampling regimen, is prohibitive. While most studies focus on the consumption of plastic, any research considering external adherence of microplastics should be aware that observed entanglement may have occurred during sampling and may be unrepresentative of microplastic–biota interactions at large. A similar interaction may occur with organisms feeding on microplastics during capture in nets; this is particularly a concern when the mesh size of the net is capable of collecting microplastics, for example, in manta nets (common mesh size 0.33 mm) [69]. Control of microplastic contamination is discussed in Section 8.2.4.

8.2.1.1.3 Sample Storage

Consideration should be given to the storage of biotic samples. Choice of preservation technique will largely depend on the research question being considered; for example, will the fixative affect the structure, microbial surface communities, chemical composition, color or analytical properties of any microplastics within the sample? Four percent formaldehyde and 70% ethanol are commonly used fixatives; however, consultation of resistance tables suggests these preservatives, albeit at higher concentrations, can damage some polymers; for example, polyamide is only partially resistant to 10% formaldehyde solution, while polystyrene can be damaged by 100% alcohol. Alternative methods for storage of organisms include desiccation [12] and freezing [41,77,83,89].

8.2.1.2 Laboratory-Exposed Organisms

Laboratory studies have been implemented to better understand the interactions between microplastics and biota. Controlled laboratory exposures facilitate monitoring of the uptake, movement and distribution of synthetic particles in whole organisms and excised tissues (e.g., gills, intestinal tract, liver). Fluorescently labeled plastics, either purchased or dyed in the lab [17], allow visualization of microplastics in organisms with transparent carapaces [10,15,30], circulatory fluids [47,49] or histological sections [105]. Where dissection is prohibitive (e.g., mussels), fluorescent microplastics can be quantified by physically homogenizing tissues followed by microscopic analysis of subsampled homogenate [35]. Coherent anti-Stokes Raman scattering (CARS) has also been used to visualize non-fluorescent nano- and microplastics in intestinal tracts and those adhered to external appendages of copepods and gill lamellae of crabs [10,35]. Bio-imaging techniques, however, are not feasible with field-sampled biota, as environmental plastics do not fluoresce and may be obscured by tissues or algal fluorescence.

8.2.2 Isolating Microplastics

In recent years, an increasing number of techniques have been developed to detect microplastics consumed by biota. Methods for extracting microplastics from biotic material include dissection, depuration, homogenization and digestion of tissues with chemicals or enzymes. Here, a range of optimized methods is consolidated, and their benefits, biases and areas of concern are evaluated.

8.2.2.1 Dissection

In a large proportion of studies, researchers target specific tissues, primarily the digestive tract (including the stomach and intestine). In larger animals, including squid [64], whales [97,98], turtles [95] and seabirds [96], dissection of the gastrointestinal tract and subsequent quantification of synthetic particles from the gut is the predominant method for assessing plastic consumption. In laboratory studies, it is more common for the whole organism (42% of studies) or the digestive tract (26% of studies) to be digested or analyzed (Figure 8.3a). In comparison, 69% of field studies targeted the digestive tract, and 27% looked at the whole organism (Figure 8.3b). Excision of the intestinal tract can also be used to ascertain consumption of microplastics by invertebrates and vertebrates including pelagic and demersal fish [19,34,41,65–67,69–80,83–93]. Investigation of stomachs and intestines is relevant for microplastics >0.5 mm in size. Microplastics larger than this do not readily pass through the gut wall without pre-existing damage, and the likelihood of translocation into tissues is too low to warrant regular investigation [131,132]. Localization of microplastics <0.5 mm can be determined by excising organs, such as the liver or gills [62,81,105] or, where the research question relates to risks of human consumption, edible tissues, for example, tail muscles of shrimp [38]. Microplastics present in dissected tissues can be isolated using saline washes, density flotation, visual inspection or digestion (see below).

8.2.2.2 Depuration

Should microplastic ingestion be the primary focus of the study, it is important that any externally adhered plastics are removed prior to treatment; typically, this is achieved by washing the study organism with water or saline water or using forceps [16,61]. A depuration step can be used to eliminate transient microplastics present in the intestinal tract. Depuration is facilitated by housing animals in microplastic-absent media (e.g., freshwater, seawater, sediment), with or without food, and leaving sufficient time for complete gut evacuation [54]; media should be refreshed regularly to prevent consumption of egested microplastics [23]. Depuration ensures only microplastics retained within tissues or entrapped in the intestinal tract are considered [23,37,51]. Depuration also provides opportunities for the collection of fecal matter, typically sampled via siphon, or pipette; feces can subsequently be digested, homogenized or directly visualized to assess and quantify egested microplastics. Fecal analysis has been used to determine microplastic consumption in a range of taxa, including sea cucumbers [28], copepods [13,17], isopods [42], amphipods [33], polychaetes [22] and mollusks [43–55].

8.2.2.3 Digestion

Enumerating microplastics present in biota, excised tissues or environmental samples can be challenging because the plastic may be masked by biological material, microbial biofilms, algae and detritus [12]. To isolate microplastics, organic matter can be digested, leaving only recalcitrant materials (Table 8.1). Traditionally, digestion is conducted using strong oxidizing agents; however, synthetic polymers can be degraded or damaged by these chemical treatments, particularly at higher temperatures. Environmentally exposed plastics, which may have been subject to weathering, abrasion and photodegradation, may have reduced structural integrity and resistance to chemicals compared to that of virgin plastics used in these stress tests [133]. As such, data ascertained using caustic digestive

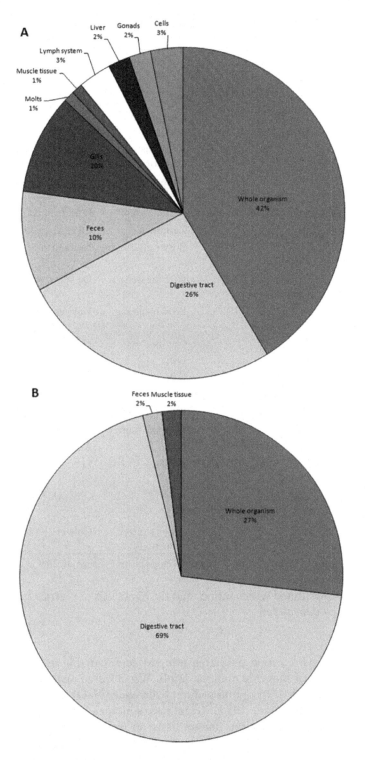

FIGURE 8.3   Target tissues of animals exposed to microplastics (a) under laboratory conditions and (b) in the environment. Total number of studies 120.

TABLE 8.1   Optimized Protocols for Digesting Biota or Biogenic Material to Isolate Microplastics

Treatment	Exposure	Organism	Reference
HNO_3 (22.5 M)	20°C (12 h) + 100°C (2 h)	Blue mussels	Claessens et al. (2013) [51]
HNO_3 (22.5 M)	20°C (12 h) + 100°C (2 h)	Blue mussels oysters	Van Cauwenberghe & Jansen (2014) [54]
HNO_3 (22.5 M)	20°C (12 h) + 100°C (2 h)	Blue mussels lugworms	Van Cauwenberghe et al. (2013) [23]
HNO_3 (100%)	20°C (30 min)	Euphausiids copepods	Desforges et al. (2015) [16]
HNO_3 (69%–71%)	90°C (4 h)	Manilla clams	Davidson & Dudas (2016) [61]
HNO_3 (70%)	2 h	Zebrafish	Lu et al. (2016) [105]
HNO_3 (22.5 M)	20°C (12 h) + 100°C (15 min)	Brown mussels	Santana et al. (2016) [63]
HNO_3 (65%) $HClO_4$ (68%) (4:1)	20°C (12 h) + 100°C (10 min)	Blue mussels	De Witte et al. (2014) [52]
HNO_3 (65%) $HClO_4$ (68%) (4:1)	20°C (12 h) + 100°C (10 min)	Brown shrimp	Devriese et al. (2015) [38]
CH_2O_2 (3%)	72 h	Corals	Hall et al. (2015) [32]
KOH (10%)	2–3 weeks	Fish	Foekema et al. (2013) [75]
KOH (10%)	60°C (12 h)	Fish	Rochman et al. (2015) [58]
KOH (10%)	2–3 weeks	Fish	Lusher et al. (2016) [89]
H_2O_2 (30%)	60°C	Blue mussels	Mathalon & Hill (2014) [53]
H_2O_2 (30%)	20°C (7 d)	Biogenic matter	Nuelle et al. (2015) [137]
H_2O_2 (15%)	55°C (3 d)	Fish	Avio et al. (2015) [81]
H_2O_2 (30%)	65°C (24 h) + 20°C (<48 h)	Bivalves	Li et al. (2015) [57]
NaClO (3%)	20°C (12 h)	Fish	Collard et al. (2015) [82]
$NaClO_3$ (10:1)	20°C (5 min)		
Proteinase K	50°C (2 h)	Zooplankton copepods	Cole et al. (2014) [12]

Assumptions: "Overnight" Given as 12 h; "Room Temperature" Given as 20°C.

agents should be interpreted with caution, and the likely loss of plastics from the digestive treatment carefully considered.

8.2.2.3.1 Nitric Acid

Nitric acid (HNO_3) is a strong oxidizing mineral acid, capable of molecular cleavage and rapid dissolution of biogenic material [134]. When tested against hydrochloric acid (HCl), hydrogen peroxide (H_2O_2) and sodium hydroxide (NaOH), HNO_3 resulted in the highest digestion efficacies, with >98% weight loss of biological tissue [51]. The optimized protocol involved digesting excised mussel tissue in 69% HNO_3 at room temperature overnight, followed by 2 h at 100°C. Desforges et al. [16] also tested HNO_3, HCl and H_2O_2 in digesting zooplankton and similarly identified nitric acid as the most effective digestion agent based on visual observations; here, the optimized digestion protocol consisted of exposing individual euphausiids to 100% HNO_3 at 80°C for 30 minutes. Adaptations of nitric acid protocols have been successfully used to isolate fibers, films and fragments from a range of organisms [23,38,54,61,64,105]. While largely efficacious in digesting organic material, a number of studies observed that oily residue and/or tissue remnants remained postdigestion [15,51,63] which have the potential to obscure microplastics. In response, De Witte et al. [52] proposed using a mixture of 65% HNO_3 and 68% perchloric acid ($HClO_4$) in a 4:1 v/v ratio (500 mL acid to 100 g tissue) to digest mussel tissue overnight at room temperature followed by 10 minutes boiling; this resulted in the removal of the oily residue. Recovery rates for 10 and 30 mm PS microspheres spiked into mussel tissue and subsequently digested with nitric acid range between 93.6% and 97.9% [51]. However, the high concentrations of acid and temperatures applied resulted in the destruction of 30×200 mm nylon fibers and melding of 10 mm polystyrene microbeads following direct exposure. Researchers have found that polymeric particles, including polyethylene (PE) and polystyrene (PS), dissolved following overnight exposure and 30 minutes boiling with 22.5 M HNO_3 [81,135]. Polyamide (PA, nylon), polyester (PET) and polycarbonate have low resistance to acids, even at low concentrations; furthermore, high concentrations of nitric, hydrofluoric, perchloric and sulfuric acid are likely to destroy or severely damage the majority of polymers tested, particularly at higher temperatures. The absence of synthetic fibers in biota digested using HNO_3 is likely a reflection of the destructive power of the acid [57].

8.2.2.3.2 Other Acids

Formic and hydrochloric acids (HCl) have also been suggested as digestive agents. With scleractinian corals (*Dipsastrea pallida*), formic acid (3%, 72 h) has been used to decalcify polyps to assist in the visualization of ingested blue polypropylene shavings [32]. HCl has also be trialed as a digestant of microplastics from pelagic and sediment samples; however this non-oxidizing acid proved inconsistent and inefficient in digesting organic material [12].

8.2.2.3.3 Alkalis

Strong bases can be used to remove biological material by hydrolyzing chemical bonds and denaturing proteins [136]. Excised fish tissues, including the esophagus, stomach and intestines, have been successfully digested using potassium hydroxide (KOH, 10%) following a 2–3 week incubation [75,89]. The protocol has been adapted for the dissolution of gastrointestinal tracts of fish and mussel, crab and oyster tissues, either directly or following baking (450°C, 6 h), by incubating tissues in 10% KOH at 60°C overnight [58,135]. This latter method has proved largely efficacious in removing biogenic material, being well suited to the dissolution of invertebrates and fish fillets, but proving less applicable for fish digestive tracts owing to the presence of inorganic materials; as with HNO_3, an oily residue and bone fragments may remain following digestion. Another strong base, sodium hydroxide (NaOH; 1 M and 10 M), has been successfully applied to remove biogenic material (e.g., zooplankton) from surface trawls, with 90% efficiency based on sample weight loss [12]. Foekema et al. [75] suggest polymers are resistant to KOH, and Dehaut et al. [135] showed no demonstrable impact on polymer mass or form, except in the case of cellulose acetate (CA). Testing the rapid KOH digestion protocol achieved a 100% microplastic recovery rate [135]. Tabulated data confirm PA, PE and polypropylene (PP) are resistant to 10% KOH, but polycarbonate (PC) and PET are degraded. Cole et al. (12 tested 40% NaOH (60°C) on a range of polymers and observed deformation of PA fibers, yellowing of polyvinyl chloride (PVC) granules and melding of polyethylene particles; similarly. Dehaut et al. [135] noted PC, CA and PET were degraded using this protocol. Notably, the compiled chemical resistance data indicate PE and PVC are resistant to NaOH, even at concentrations of 50% at 50°C. That Cole et al. [12] observed changes in polymers supposedly resistant to the given treatment highlights the necessity for comprehensive testing of applied treatments prior to use on biota.

8.2.2.3.4 Oxidizing Agents

Hydrogen peroxide (H_2O_2) and peroxodisulfate potassium ($K_2S_2O_8$) are oxidizing agents. Mathalon and Hill [53] used H_2O_2 (30%) at 55°C–65°C to digest mussel soft tissue, and although largely effective, the authors noted "flakes of debris" remained. Li et al. [57] also applied this method, but incubated samples in an oscillating incubator, and then at room temperature for 24–48 h. Avio et al. [81] similarly tested alternate treatments in digesting intestinal tracts of mullet (*Mugil cephalus*); while H_2O_2 was an efficacious digestant, Avio et al. [81] identified that direct application of H_2O_2 resulted in only a 70% retrieval of spiked microplastics, with losses linked to H_2O_2 foaming. A number of studies noted excessive foaming might obscure samples or lead to sample loss [51,53,81]. A density separation of stomach contents with hypersaline (NaCl) solution followed by digestion of isolated material with 15% H_2O_2 resulted in a much improved 95% recovery rate for spiked microplastics. Dehaut et al. [135] trialed 0.27 M $K_2S_2O_8$ with 0.24 M NaOH in digesting biological tissues; while largely efficient in digesting biogenic material (<99.7% mass reduction), the authors noted its expense and highlighted issues with crystallization of the digestive solutions and incomplete digestion causing blockages during filtration. Avio et al. [81] observed that 15% H_2O_2 had no visible impact on PE or PS microspheres, although slight modifications to FTIR spectra were observed. Conversely, Nuelle et al. [137] identified some visual deformities to exposed plastic and quantified a 6.2% loss in size for PP and PE particles (<1 mm). Resistance data indicate 30% H_2O_2 should have little or no effect on PE or PP, again highlighting the importance of thorough testing of protocol applicability. Tabulated data indicate PA and PE are also prone to damage or dissolution from 30% H_2O_2. $K_2S_2O_8$ resulted in no changes in the mass or appearance in the majority of exposed polymers, but caused complete dissolution of cellulose acetate [135]; chemical resistance data are currently unavailable, and this requires further testing.

8.2.2.3.5 Sodium Hypochlorite

Sodium hypochlorite (NaClO) is used as an endodontic irrigant, with a near linear dose-dependent dissolution efficiency for biological tissue [138]. Collard et al. [82] digested fish stomach contents, in an overnight exposure with ~3% NaClO; filtered digestants were subsequently washed with 65% HNO_3 and digested in a 10:1 NaClO and HNO_3 solution for 5 minutes. The technique caused no visible degradation of a range of polymers (PET, PVC, PE, PP, PS, PC or PA). Resistance chart data indicate 15% NaClO would degrade PA, although no data are provided for the ~3% NaClO concentration applied by Collard et al. [82].

8.2.2.3.6 Enzymes

Enzymatic digestion has been mooted as a biologically specific means of hydrolyzing proteins and breaking down tissues [12]. To remove biological material from field-collected samples, Cole et al. [12] developed a digestion protocol employing a serine protease (Proteinase K). Material was desiccated (60°C, 24 h), ground and homogenized by repeatedly drawing samples through a syringe. Next, samples were mixed with homogenizing solution (400 mM Tris–HCl buffer, 60 mM EDTA, 105 mM NaCl, 1% SDS), acclimated to 50°C, enzymatically digested with Proteinase K (500 mg/mL per

0.2 g DW sample) and mixed with sodium perchlorate ($NaClO_4$, 5 M). Ultrasonication was demonstrated to have a deleterious effect on digestion efficiency, owing to protein precipitation in the media. With marine samples, the Proteinase K method proved to have a digestion efficacy of >97%, and the method was used to isolate fluorescent polystyrene microspheres (20 mm) ingested by marine copepods. The authors note that additional enzymes could be used depending on the chemical makeup of the organism or samples in question (e.g., chitinase with chitinous invertebrates). The enzyme pepsin causes no damage to polymers, but proved only partially effective at digesting biogenic material [137]. It has recently been reported that enzymes have been successfully applied in the isolation of microplastics from: intestinal tracts of turtles with Proteinase K [139], mussel tissue with Corolase 7089 (AB enzymes) [140] and herring digestive tracts with Proteinase K and H_2O_2 [141]. In contrast to chemical digestion techniques, enzymes ensure no loss, degradation or surface change to plastics present [12] and are less hazardous to human health. The trade-off is a protracted method, necessitating increased researcher time when considering large-scale field sampling and monitoring.

8.2.2.3.7 Filtering Digestants
Following digestion, chemical agents can be filtered to retain any recalcitrant materials (e.g.. undigested tissue, inorganic residue, microplastics). Viable filters include 0.2 and 0.7 mm glass fiber filters [61,63], 5 mm cellulose nitrate filters [54], 5 mm cellulose acetate membranes [82], 50 mm mesh [12] and 250 mm mesh [77]. Larger pore size facilitates rapid filtering but will result in the loss of smaller plastics [12]. Glass fiber filters can shed and might be considered a source of contamination; smoother filters (e.g., membrane filters) are typically easier to scrape and less prone to fragment (personal observations of the authors). Microplastics on filters can be visualized directly (see Section 8.2.3), transferred to slides [63] or extracted. Collard et al. [82] suggest placing filters in methanol solution, ultrasonicating (50 Hz), centrifuging (5000 rpm, 5 min, 20°C) and then removing pelleted plastic by pipette; while this method was suitable for a range of polymers, the methanol caused a 25% weight reduction in tested PVC particles.

8.2.2.4 Density Separation
Although most commonly utilized in studies of water and sediment samples, density separation has been used in four biotic studies. Three studies used NaCl to separate less-dense particles [53,57,81] while Collard et al. [82] used a centrifuge. Following settlement of denser materials, the supernatant is filtered and the resulting material examined under a microscope. Density separation can be useful in studies following digestion. Saturated salt solutions, such as NaCl (aq), allow the separation of less-dense particles where there are large amounts of inorganic matter (e.g., sand, chitin, bone) that have not been dissolved (Lusher, personal observations). Density separation has been recommended by the MSFD (EU) for Europe. NaCl is recommended because it is inexpensive and nonhazardous; however, the use of NaCl could lead to an underestimation of more-dense particles (>1.2 g/cm³). NaI and $ZnCl_2$ solutions have been considered as viable alternatives to NaCl (aq) [142]. Their high density makes them capable of floating high-density plastics, including PVC.

8.2.3 Microplastic Identification

Following the preparation of target tissues, the quantity and types of microplastics should be ascertained. Of the methods currently employed, visual identification is most widely utilized, often in combination with one or more follow-up analytical techniques. Researchers can use characteristics, including morphology and density, to identify the presence of microplastics. Visual identification is based on the morphological and physical characteristics of particles whereas chemical characteristics are determined by more advanced analytical techniques.

8.2.3.1 Visual Identification

Early reports quantifying environmental plastics primarily relied upon visual identification; this method remains an essential step in classifying microplastics and is perfectly acceptable when supported by subsequent polymer analysis of subsamples. Visual identification can be conducted using light, polarized or electron microscopy. Semi-automated methods, including ZooScan [143], flow cytometry [144], cell sorters and coulter counters [14] allow for a large number of samples to be analyzed rapidly; however, these require technical expertise and specialized equipment, and time must still be given to sample preparation and data analysis. Scanning electron microscopes (SEM) produce high-resolution images and have been implemented in several studies either to image recovered plastics [23,41] or as a way of identifying microbial colonization [145].

Visual identification is rapid, relatively cheap and can be conducted without the need for additional technical staff and consumables; however, accurately differentiating microplastics, particularly in the smaller size ranges, requires training and experience. Criteria for visually identifying microplastics include the absence of cellular or organic structures, a homogenous thickness across the particles and homogenous colors and gloss [77,123]. Manually sorting plastics under a microscope is most effective for particles >500 μm; the effort and accuracy required for sorting increases with decreasing particle size. Owing to the difficulties in handling and differentiating microplastics from organic and inorganic matter [146], error rates could be as high as 70%, increasing with decreasing particle size [61], with incorrect identification most prevalent with microfibers [123,147]. To gauge the accuracy of visual discrimination, subsamples of potential plastics should be chemically analyzed [77,123,147–150]. It has been observed that training and experience can significantly lower the error rates and misidentification stemming from visual identification [83].

Plastics are largely classified by their morphological characteristics: size, shape, and color. Size is typically based on the longest dimension of a particle; size categories can be used where appropriate. When reporting microplastic shape, researchers tend to use five main categories, although the nomenclature used varies between research groups (Table 8.2). Finally, colors are often reported across a wide spectrum; color differentiation is subjective and visual identification of microplastics cannot be based on color alone. Caution should be given to categorizing microplastics suffering embrittlement, fragmentation or bleaching, or encrusted with biota, as this may skew results.

TABLE 8.2   Categories Used when Classifying Microplastic by Shape

Shape Classification	Other Terms Used
Fragments	Irregular shaped particles, crystals, fluff powder, granules, shavings
Fibres	Filaments, microfibers, strands, threads
Beads	Grains, spherical microbeads, microspheres
Foams	Polystyrene, EPS
Pellets	Resin pellets, nurdles, pre-production pellets, nibs

8.2.3.2 Polymer Verification

Due to the challenges in visually identifying microplastics, secondary analyses should be used to confirm the identity of suspected polymeric material. The method employed is often dictated by the equipment available and while any chemical characterization of the polymers recovered is useful, some techniques are more robust than others. The European Commission suggests that a subsample (5%–10%) of particles with a size between 100 mm and 5 mm and all particles between 20 and 100 mm should be subjected to further verification techniques. Postvisual analyses have shown misidentification of microplastics in wild-caught animals of up to 70% [82,147,150]. It should be noted here that errors in identification often include unmatched spectra that could not be assigned with confidence to a known polymer type; confidence thresholds for spectra matches are usually set at 70%–75% [77,83,89]. Confirming the identity of suspected plastics may be carried out in a number of ways depending on the funds and equipment available to the researcher. Perhaps the simplest technique is the use of a hot needle to observe melting points [38,52,60,86]. While both cheap and fast, this method does not allow for the accurate identification of the polymer; however, the temperature range at which melting occurs does provide a specific range of potential plastics. A converse method is to exclude nonplastics rather than identifying the plastics present; oven and freeze drying removes water from organic material causing it to wither. This increases the likelihood of nonplastic material being identified and removed from mixed samples [151,152]. Combining these two techniques provides a cheap, if laborious, method of plastic identification.

Another low-cost technique involves the examination of microplastics under a polarized light microscope to observe the birefringent properties of the suspected polymer. The birefringence of a polymer is the result of its chemical structure and manufacturing methods which results in unique anisotropic properties; by passing polarized light through a sample, unique spectra are created, from which it is possible to confirm the identity of plastic materials [153]. As with the hot needle technique, this method require plastics to be viewed individually; while initial costs are low, the time taken makes it prohibitive for large samples. More complex—and costly—methods can also be used to infer resin constituents, plastic additives and dyes.

Often, these techniques require the purification of the potential microplastic prior to analysis. The removal of biofilms and organic and inorganic matter adhered to the surface will avoid impeding polymer identification and the removal of nonplastic particles [146]. Following purification, suspected plastics are submitted to analytical techniques including Fourier-transformed infrared spectrometry (FTIR) in transmittance or reflectance; attenuated total reflectance (ATR); Raman spectrometry for color pigment spectra and pyrolysis-gas chromatography combined with mass spectroscopy (Pyr-GC-MS), which analyzes particles using their thermal degradation properties and can be used to analyze polymer type and organic plastic additives simultaneously [154]. Alternate analytical methods include high-temperature gel-permeation chromatography (HT-GPC) with IR detection; SEM-EDS and thermo-extraction and desorption coupled with GC-MS [150,155,156].

If coupled with microscopy, FTIR and Raman can be used to identify microplastics with a size >20 mm [123,149]. Raman spectroscopy combined with microscopy has a higher resolution (approx. 1–2 mm) [100,149] and can be used to locate particles within biological tissues [10]. FTIR and Raman have been recommended for determining resin constituents [123,149]. There is minimal sample preparation, other than clean up, required for FTIR. However, FTIR and Pyr-GS-MS are both destructive. Raman is nondestructive as it does not require the sample to be flattened or manipulated. The disadvantages of Pyr-GS-MS is the manual placement of the particle in the instrument, which can incur size limitations and only one particle can be run per sample. However, qualitative and quantitative analyses are being developed [141,157]. A drawback of chemical analysis is that the isolation of small, highly degraded samples increases the chances of misidentification and producing noisy spectra in which the vital fingerprint areas are obscured, although this can be improved by the use of microscope-aided instrumentation (micro-FTIR and micro-Raman), which is designed to target and read responses from samples of a smaller size.

8.2.4 Contamination

At all stages, care must be taken to prevent the contamination or cross-contamination of samples. Airborne contamination of samples with synthetic fibers stemming from clothing [75,158] or atmospheric fallout [159] is a recurrent issue within the literature [52,53,58,61,63]. Sources of contamination should be eliminated where possible and otherwise quantified using environmental filters or procedural blanks. Here, we highlight sources of contamination during sampling and sample processing and consider protocols for contamination mitigation.

8.2.4.1 Contamination during Field Sampling

With marine species, animals are often sampled by way of polymer rope, nets or traps [123]. In these situations, animals should only be exposed for minimal periods and a reference sample of the gear should be retained to exclude contamination during the identification phase [77]. Avoiding airborne contamination of samples in the field is understandably more complex than in the laboratory, but remains an important consideration, nevertheless. Steps for mitigating contamination include thorough cleaning of all equipment prior to sampling, which will also mitigate cross-contamination; covering samples and equipment between use; wearing polymer-free clothing or cotton coveralls, and gloves and the use of procedural blanks.

8.2.4.2 Contamination during Sample Processing and Analysis

In the laboratory, forensic techniques, good laboratory practice and common sense should be applied to mitigate contamination [160]. Wherever feasible, researchers should process samples in a laminar flow hood (e.g., cell or algal culture unit) [12,54,55,82]; alternatively a fume hood [63] or "clean room" (i.e., nonventilated or negative flow) with low foot-traffic or embargoed to non-essential personnel can be used. Glassware is preferential to plastic consumables; Cole et al. [12] observed physical homogenization of specimens in polypropylene Falcon tubes resulted in the introduction of plastic shavings to the sample. Filtering media or liquids used in sample preparation has been recommended by some researchers [54,55]. Glassware, benches and equipment should be rinsed with deionized water [53–55,58,63,77,89], ethanol [12,82] or acetone [52] prior to use. Collard et al. [82] further suggests drying equipment with cellulose-lignin-based cloths from which reference samples can be taken. As with field sampling, all materials should be covered between use, and cotton coveralls or laboratory coats are widely recommended. Environmental filters (e.g., glass fiber filters) can be placed near equipment to quantify external contamination [77,89]. Lastly, procedural blanks (i.e., controls) are highly recommended for quantifying contamination and for identifying aspects of the experimental design where contamination can occur. Analysis of procedural blanks can reveal substantial contamination of synthetic fibers, ranging 5.8 ± 2.2 [61] to 33–39 fibers [137] per replicate. Where contaminating plastics are easily identifiable, for example, being brightly colored [137], >1.5 mm [38] or <36 mm [82], or resembling laboratory coat fibers [58], these microplastics can be removed from subsequent analysis.

8.2.5 Data Analysis

The varied methods by which microplastic uptake by biota is measured understandably result in differing levels of recording. At the highest level, researchers record the number of items, often in relation to organism size. This may be recorded simply, as the percentage of individuals seen to ingest microplastic [36,66,71,76,77,79,84,89], the number of microplastics per individual [16,53,54,57,58,65,74,78,80,85,88–91,102] or as the number of microplastic items by length or weight [38,54,60,61,73,77,89]. Many types of plastic, for example, microfiber boluses, do not lend themselves to the enumeration of individual plastic items. In addition, mastication and peristaltic action may break down plastic items within the gut; as a result, the number of items in the gut may exceed that originally ingested. In such cases, researchers have reported the weight of plastic aggregations [41], descriptions of the aggregation of microplastic observed [34] or a combination of the two. Such issues in enumeration are more often observed in studies of wild-caught organisms, where the initial level of microplastic exposure is not known and the type of microplastic recovered is susceptible to tangling. A similar issue may be observed in the study of microplastic uptake in laboratory experiments; here, concentrations of introduced microplastic may be recorded solely by number or mass per individual [19,161] or as a value in relation to mass of food [33,34,92,99,161–163] or volume of water [10,26,164].

The use of multiple methods to quantify the level of microplastic uptake by fish and invertebrates is also an issue in the reporting of environmental plastic levels. Inconsistency in the use of units can mask or inflate the apparent impact of microplastics on a species or location. This increases the likelihood of errors arising when comparing multiple studies carried out by unrelated researchers. The manner in which plastic abundance and concentration is recorded influences the range of statistical analyses available; for example, grouping aggregations into specific classifications reduces the power of the available tests. In field experiments, a range of techniques have been used to determine the relationship between microplastic uptake and both biological and environmental factors. Many of these methods combine continuous and categorical variables in linear models of varying complexity and require careful structuring in statistical software such as R statistical software [165].

8.3 DISCUSSION

Techniques used to isolate and enumerate microplastics in fish and invertebrates have largely been adapted from studies focused on large vertebrates, or have derived from traditional biology methods; as these fields advance, it is vital that we continue to observe their progress and incorporate relevant methods. Owing to the challenges in sampling, isolating and identifying microplastics, and the diverse physiology of taxonomic groups under investigation, a degree of flexibility, innovation and ingenuity on the part of the researcher is clearly required.

Of the numerous studies investigating microplastic uptake in fish and invertebrates, it is analysis of wild-caught animals that presents the most issues. In these studies, potential sources of error are numerous, including microplastic losses or contamination during sampling; furthermore, there are substantial challenges in drawing links between exposure and effect. These studies, however, are essential for establishing ecologically relevant data, which ultimately provides researchers with a clear view of the quantity and types of plastic (and associated contaminants) experienced by biota in the natural environment. In this section, we address the need for standardizing protocols for microplastic quantification

[166], outlining preferred methods, best practice and steps for mitigating contamination. It is anticipated that compiling standardized methodologies will provide researchers with a grounding in developing future experimental design.

8.3.1 Controlling Sampling Bias

Throughout the sampling process, utmost care must be taken to prevent the artificial inflation or loss of microplastics. In Table 8.3, the commonly used methods for sampling and isolating microplastics across a range of taxa were outlined. In all cases, the least damaging sampling gear is preferable, and sampling periods should be kept as short as practically realistic. Organisms that spend longer in nets are subject to additional stress that increases the likelihood of regurgitation or stomach inversion and artificially increases contact time between microplastics and biota; this could facilitate microplastic ingestion and adherence to external appendages. Individuals should be rinsed following capture to remove adhered particles, and samples of fishing gear should be taken to

TABLE 8.3　Standard Sampling and Plastic Isolation Strategies Employed across a Variety of Subphyla

Ecosystem	Life Strategy	Subphylum	Size Range	Sampling Method	Initial Plastic Separation
Aquatic	Benthic	Annelida	—	Grabs	Digestion
Aquatic	Benthic	Crustacea	>50 mm	Trawls/creels	Dissection
Aquatic	Benthic	Crustacea	<50 mm	Otter-/beam-trawls	Digestion
Aquatic	Benthic	Echindodermata	—	Grab/trawls	Dissection
Aquatic	Benthic	Mollusca	>30 mm	Grabs	Dissection
Aquatic	Benthic	Mollusca	<30 mm	Grabs	Digestion
Aquatic	Benthic	Flatftsh	—	Otter-/beam-trawls	Dissection
Aquatic	Nektonic	Crustacea (juv.)	<50 mm	Midwater trawls	Digestion
Aquatic	Nektonic	Gadids		Otter-/midwater trawls	Dissection
Aquatic	Nektonic	Echindodermata (juv.)	<2 cm	Trawls	Digestion
Aquatic	Nektonic	Mollusca (juv.)	<2 cm	Trawls	Digestion
Aquatic	Nektonic	Fish	<10 cm	Midwater trawls	Dissection
Aquatic	Planktonic	Annelida	<2 cm	Trawls	Digestion
Aquatic	Planktonic	Cnidaria	<10 cm	Trawls	Digestion
Aquatic	Planktonic	Crustacea	<2 cm	Trawls	Dissection
Terrestrial	—	Annelida	—	Sediment collection	Digestion
Terrestrial	—	Arachnida	—	Trapping/hand gathering	Digestion
Terrestrial	—	Crustacea	—	Trapping/hand gathering	Digestion
Terrestrial	—	Insecta	—	Trapping/hand gathering	Digestion
Terrestrial	—	Mollusca	—	Trapping/hand gathering	Digestion

exclude material ingested as a result of capture. Researchers are recommended to over-sample where practical, so that individuals with recently emptied stomachs, or otherwise damaged during sampling, should be omitted from the dataset; this will help reduce the bias caused by regurgitation and enable more robust comparisons among animals sampled from different sites or collected using alternate sampling methods. Specimens should be rapidly transported to the laboratory or preserved promptly to avoid microplastic egestion in transit. The collection from commercial fish markets or artisanal fishers is not ideal, as the researcher will have less, if any, control on the method of capture and the handling conditions on transport. Where applicable, researchers are suggested to work closely with fishers to ensure animals are sampled appropriately and sufficient information on the capture procedure is collected.

8.3.2 Effective Plastic Isolation

Researchers are presented with a range of techniques for isolating microplastics from biota, including dissection, depuration, digestion and density separation. Determining the appropriate method will largely depend on the research question (e.g., risks of human consumption, total body burden, localized accumulation). Digestion of whole organisms or excised tissues is widely used (Table 8.2), however caution must be given in selecting an appropriate digestive agent because of the potential destruction of contaminants. For example, >50% formic acid, >35% $HClO_4$, >40% hydrofluoric acid, >80% H_2O_2, >50% HNO_3, >70% $HClO_4$, >50% KOH and >95% sulfuric acid can be particularly damaging to specific polymers. Some digestive agents, including H_2O_2 and $HClO_4$, are simply ineffective in breaking down tissues. Our analysis found that some recent studies used high percentages of acids to which several polymers are not resistant. In particular, high percentages of $HClO_4$ have been utilized in several studies [16,23,38,51,52,54,60,61,105]. For example, only PET and PVC are resistant to 50% HNO_3, whereas, PE and PP are partially resistant and PA, PC and PS are not resistant. Typically, a balance will need to be struck between finding a cost-effective digestive agent with the capacity to effectively break down tissue without losing microplastics. Based on our analyses, the rapid 10% KOH (60°C overnight) [58,135] and enzymatic digestion protocols [12,140] appear, on balance, to be among the most widely tested and effective digestive treatments currently available; in all cases, the costs, strengths and weaknesses, and applicability of each method to the study organism in question, should be carefully considered. As with sampling, steps for mitigating and accounting for external contamination are paramount.

8.3.3 Polymer Verification

Methods for verifying isolated microplastics vary in complexity and expense. The method used is dependent not only on resources available to the research group, but also the degree of information required by the study. Studies examining the total body burden or the rate of uptake may only require the most cursory identification to confirm that the particles recovered are indeed plastic, whereas research examining the potential origins of plastics or the presence of adsorbed contaminants and additives require more rigorous testing. We concur with the European standard of polymeric identification of a 5%–10% subsample of isolated microplastics. While it has been suggested that this level is insufficient to accurately determine the ability of the researcher to accurately identify plastics, this can be improved when selecting subsamples for verification. Researchers must ensure a representative sample encompassing all categories of recovered microplastics, and particular consideration should be given to commonly misidentified forms, such as fibers and small size fractions. Where possible, researchers should include the weight and number of plastic items, and the masses of sampled organisms should be recorded in cases where animals are grouped prior to analysis. Such standardization will improve comparability between studies.

8.3.4 Mitigating Contamination

Contamination, cross-contamination and loss of plastics are a major challenge for microplastics research. It is recommended that all laboratory processing should include steps for preventing or limiting airborne contamination, and procedural blanks used to account for this [124]. Additional processing may be utilized during microplastic isolation to improve the detectability or identification of microplastics; however, each additional step increases the opportunity for contamination. Where possible, plastic consumables should be avoided [167]. All samples should be preserved by freezing, desiccation or in filtered ethanol or formalin, although the latter may result in loss of some plastics. On research vessels, glassware may not be feasible, so plastic may be used following sufficient cleaning with filtered water.

8.3.5 Data Analysis

Researchers should be aware of biases in sampling environmentally exposed individuals. Firstly, the condition of organisms prior to capture is unknown, and linking microplastic (and co-contaminant) burden with condition is prohibitive. Secondly, sampling may lead to an underestimation of the microplastic burden in a population because highly contaminated individuals are dead or dying, or remain in shelters and burrows owing to reduced functionality. Thirdly, it is vital that sampling is spatially and/or temporally broad to ensure that observed levels of contamination are representative of the wider population. For example, Welden et al. [41] observed significant spatial variation in fiber contamination in three Scottish populations of *Nephrops norvegicus*.

8.3.6 Recommendations for Future Work

In reviewing the relevant literature, it is apparent that research is currently skewed toward vertebrates (Figure 8.4). The range of ecological functions carried out by invertebrates and the diverse niches which they occupy all suggest that the impacts of microplastics on these groups may have a marked effect on the environment. Further assessment is recommended of the uptake and impact of microplastics in these groups as this is essential if we are to predict the extent of the effects on biodiversity, ecosystems and ecological processes. A comparison of the relative uptake and retention of the different categories and shapes of microplastics is also required to determine which are the most harmful. Many laboratory studies of microplastic ingestion rely solely on pre-produced plastics, which are easily purchased from suppliers (e.g., polystyrene microbeads [10–15,17,26,47,105,117]); however, these are not representative of the diverse forms currently present in the

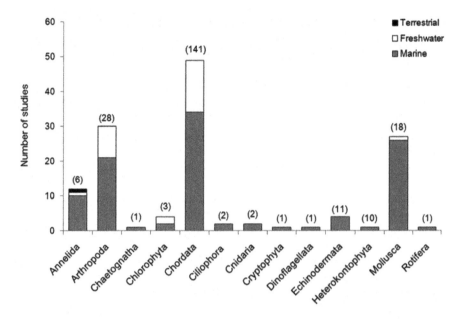

FIGURE 8.4   Laboratory and field studies investigating microplastic interactions with biota separated by Phyla. The value in parentheses above each bar indicates the number of species studied within the taxa. Total number of studies included in this review 120.

environment. Understanding which plastics are readily retained is necessary to determine the threat posed by the relative levels of environmental microplastics and to link these to evidence of negative effects in physiology and behavior, which may impair function at the ecosystem level.

In fact, a number of studies have considered the observable impacts of microplastics on organisms. In 11/120 studies reviewed here, researchers examined the relationship between microplastic uptake and changes in physiology and body chemistry. Endpoints have included organism behavior [104,168], lysosomal response [56], lipid content [169,170], protein content [170], population fitness [161], cellular population growth [171] and individual growth [15,170]. They also utilized ecotoxicological assays to monitor embryonic development [31,172] and the uptake of metals [94,173] and chemicals [162,163,174]. Laboratory studies must use environmentally realistic concentrations of microparticles to allow evaluation of potential harm to the individual as a result of field exposure. In wild populations, where the presence of confounding factors makes it difficult to attribute biological responses and condition directly to plastic exposure, direct observation of plastic type and abundance remains the most reliable method of determining microplastic impacts.

Microplastics may be selectively or nonselectively ingested or acquired through trophic transfer. Again, the artificially inflated levels of nonrepresentative plastics used in a number of currently available studies greatly increases the potential for chance transfer of plastics. The results of such studies, while useful in indicating the potential for transfer, display transfer in a way not feasible at normal contamination levels. In studies not focusing on the ecotoxicological impacts, we recommend inflated levels of microplastic should only be used in the presence of an ecologically valid control determined by reference to rigorous environmental sampling.

Lastly, few studies have addressed the movement of microplastics within ecosystems. Many rely on seeding plastic into non-natural food sources, in mesocosm experiments with no alternate food sources, and while these experiments have shown the transfer of plastic between food and organisms, there is a clear need for more robust information on the validity of these results *in natura* [175].

The need for rapid, accurate assessment of the levels of microplastic in wild populations is essential for determining baseline levels of contamination and assessing the risk of microplastics to organisms and ecosystems. The diverse physiology of the organisms covered in this review has necessitated the analysis of a range of protocols for microplastic extraction and enumeration which could limit comparability among studies. As such, we recommend the development of a standard methodology per subphylum or class of organism which will combine efficiency in digestion and recovery of microplastics with the use of the least toxic chemicals to preserve plastic polymers.

Particular attention should be given to harmonizing the way in which data are recorded (e.g., mass and number of isolated microplastics per mass of organism) to promote comparability. Prevention of overestimation of plastic contamination must be controlled by confirming the identity of a proportion of the recovered materials. This may be carried out either by chemical analysis, density separation, birefringent microscopy or other physical examination.

8.4 CONCLUDING REMARKS

In this review, various methods for sampling, isolating and identifying microplastics ingested by fish and invertebrates were examined. As research progresses, the need for method standardization becomes clear, so that a finer picture of the threat of microplastics to organisms emerges and ecological and environmental risks can be assessed. Such standardized methodologies must take into account the numerous potential sources of error and contamination, as outlined above, and also the general need for monitoring, which demands that a great number of samples be processed hastily.

Many of the studies covered in this review have focused on the issue of whether microplastic uptake occurs, only addresses the level of plastic contamination in a single species or group and do not allow the assessment of risk and disturbance at the ecosystem level. These studies have resulted in numerous further questions regarding the uptake and transfer of microplastics within ecosystems: Is plastic uptake selective, or passive? To what extent does trophic transfer occur? And, are the negative effects of plastic ingestion observed in laboratory experiments valid in the environment? There is a strong need to design studies in such a way that their results contribute to clarify these issues, for example, contrasting microplastic loads with environmental contamination or diet, to give a more holistic approach to the study of microplastic pollution.

REFERENCES

1. Plastics the Facts 2014/2015, https://issuu.com/plasticseuropeebook/docs/final_plastics_the_facts_2014_19122 (accessed August 2018).
2. Cole M, Lindeque P, Halsband C, and Galloway TS, Microplastics as contaminants in the marine environment: A review. *Mar. Pollut. Bull* 2011, 62(12), 2588–2597.
3. UNEP, *Marine Plastic Debris and Microplastic Technical Report*, United Nations Environmental Programme, Nairobi, 2016.

4. Lusher A. In: Bergmann M, Gutow L, and Klages M (eds) *Marine Anthropogenic Litter*. Springer, Berlin. 2015, pp. 245–308.
5. Wagner M, Scherer C, Alvarez-Munoz D, Brennholt N, Bourrain X, Buchinger S et al. Microplastics in freshwater ecosystems: What we know and what we need to know. *Environ. Sci. Eur* 2014, 26(12), 1–9.
6. Eerkes-Medrano D, Thompson RC, and Aldridge DC. Microplastics in freshwater systems: A review of the emerging threats, identification of knowledge gaps and prioritisation of research needs. *Water Res* 2015, 75, 63–82.
7. Duis K, and Coors A. Microplastics in the aquatic and terrestrial environment: Sources (with a specific focus on personal care products), fate and effects. *Environ. Sci. Eur* 2016, 28(1), 1–25.
8. Christaki U, Dolan JR, Pelegri S, and Rassoulozadegan F. Sampling, isolating and identifying microplastics ingested by fish and invertebrates. *Limnol. Oceanogr* 1998, 43(3), 458–464.
9. Wilson DS, Food size selection among Copepods. *Ecology*, 1973, 54(4), 909–914.
10. Cole M, Lindeque P, Fileman E, Halsband C, Goodhead R, Moger J, and Galloway TS. Microplastic Ingestion by Zooplankton. *Environ. Sci. Technol* 2013, 47(12), 6646–6655.
11. Lee K-W, Shim WJ, Kwon OY, and Kang J-H. Size-dependent effects of micro polystyrene particles in the marine copepod Tigriopus japonicas. *Environ. Sci. Technol* 2013, 47(19), 11278–11283.
12. Cole M, Webb H, Lindeque PK, Fileman ES, Halsband C, and Galloway TS. Isolation of microplastics in biota-rich seawater samples and marine organisms. *Sci. Rep* 2014, 4, 4528, DOI: 10.1038/srep04528.
13. Setälä O, Flemming-Lehtinen V, and Lehtiniemi M. Ingestion and transfer of microplastics in the planktonic food web. *Environ. Pollut* 2014, 185, 77–83.
14. Cole M, Lindeque H, Fileman ES, Halsband C, and Galloway TS. The Impact of Polystyrene Microplastics on Feeding, Function and Fecundity in the Marine Copepod Calanus helgolandicus. *Environ. Sci. Technol* 2015, 49(2), 1130–1137.
15. Cole M, and Galloway TS. Ingestion of Nanoplastics and Microplastics by Pacific Oyster Larvae. *Environ. Sci. Technol* 2015, 49(24), 14625–14632.
16. Desforges, JPW Galbraith M, and Ross PS. Ingestion of Microplastics by Zooplankton in the Northeast Pacific Ocean. *Arch. Environ. Contam. Toxicol* 2015, 69(3), 320–330.
17. Cole M, Lindeque P, Fileman ES, Halsband C, and Galloway TS. Microplastics Alter the Properties and Sinking Rates of Zooplankton Faecal Pellets. *Environ. Sci. Technol* 2016, 50(6), 3239– 3246.
18. Bolton TF, and Havenhand JN. Physiological versus viscosity-induced effects of an acute reduction in water temperature on microsphere ingestion by trochophore larvae of the serpulid polychaete Galeolaria caespitosa. *J. Plankton Res* 1998, 20(11), 2153–2164.
19. Thompson RC, Olsen Y, Mitchell RP, Davis A, Rowland SJ, John AW, McGonigle D, and Russell AE. Lost at Sea: Where Is All the Plastic? *Science*, 2004, 304(5672), 838.
20. Besseling E, Wegner A, Foekema EM, Van Den Heuvel-Greve MJ, and Koelmans AA. Effects of Microplastic on Fitness and PCB Bioaccumulation by the Lugworm Arenicola marina (L.). *Environ. Sci. Technol* 2013, 47(1), 593–600.
21. Browne MA, Niven SJ, Galloway TS, Rowland SJ, and Thompson RC. Microplastic Moves Pollutants and Additives to Worms, Reducing Functions Linked to Health and Biodiversity. *Curr. Biol* 2013, 23(23), 2388–2392.
22. Wright SL, Rowe D, Thompson RC, and Galloway TS. Microplastic ingestion decreases energy reserves in marine worms. *Curr. Biol* 2013, 23(23), R1031–R1033.

23. Van Cauwenberghe L, Claessens M, Vandegehuchte MB, and Janssen CR. Microplastics are taken up by mussels (Mytilus edulis) and lugworms (Arenicola marina) living in natural habitats.*Environ. Pollut* 2015, 199, 10–17.

24. Green DS, Boots B, Sigwart J, Jiang S, and Rocha C. Effects of conventional and biodegradable microplastics on a marine ecosystem engineer (Arenicola marina) and sediment nutrient cycling. *Environ. Pollut* 2016, 208, 426–434.

25. Gusmão F, Di Domenico M, Amaral ACZ, Martinez A, Gonzalez BC, Worsaae K, do Sul JAI, and da Cunha Lana P. In situ ingestion of microfibres by meiofauna from sandy beaches. *Environ. Pollut* 2016, 216, 584–590.

26. Setälä O, Norkko J, and Lehtiniemi M. Feeding type affects microplastic ingestion in a coastal invertebrate community. *Mar. Pollut. Bull* 2016, 102(1), 95–101.

27. Hart MW. Particle Captures and the Method of Suspension Feeding by Echinoderm Larvae. *Biol. Bull* 1991, 180(1), 12–27.

28. Graham ER and Thompson JT. Deposit- and suspension-feeding sea cucumbers (Echinodermata) ingest plastic fragments. *J. Exp. Mar. Biol. Ecol* 2009, 368(1), 22–29.

29. Della Torre C, Beergami E, Salvati A, Faleri C, Cirino P, Dawson KA, and Corsi I. Accumulation and Embryotoxicity of Polystyrene Nanoparticles at Early Stage of Development of Sea Urchin Embryos Paracentrotus lividus. *Environ. Sci. Technol* 2014, 48(20), 12302–12311.

30. Kaposi KL, Mos B, Kelaher BP, and Dworjanyn SA. Ingestion of Microplastic Has Limited Impact on a Marine Larva. *Environ. Sci. Technol* 2014, 48(3), 1638–1645.

31. Nobre CR, Santana MFM, Maluf A, Cortez FS, Cesar A, Pereira CDS, and Turra A. Assessment of microplastic toxicity to embryonic development of the sea urchin Lytechinus variegatus(Echinodermata: Echinoidea). *Mar. Pollut. Bull* 2015, 92(1–2), 99–104.

32. Hall NM, Berry KLE, Rintoul L, and Hoogenboom MO. Microplastic ingestion by scleractinian corals. *Mar. Biol* 2015, 162(3), 725–732.

33. Ugolini A, Ungherese G, Ciofini M, Lapucci A, and Camaiti M. Microplastic debris in sandhoppers. *Estuarine, Coastal Shelf Sci* 2013, 129, 19–22

34. Murray F, and Cowie PR. Plastic contamination in the decapod crustacean Nephrops norvegicus (Linnaeus, 1758). *Mar. Pollut. Bull* 2011, 62(6), 1207–1217.

35. Watts AJ, Lewis C, Goodhead RM, Beckett DJ, Moger J, Tyler C, and Galloway TS. Uptake and Retention of Microplastics by the Shore Crab Carcinus maenas. *Environ. Sci. Technol* 2014, 48(15), 8823–8830.

36. Brennecke D, Ferreira EC, Costa TMM, Appel D, da Gama BAP, and Lenz M, Ingested microplastics (>100 μm) are translocated to organs of the tropical fiddler crab Uca rapax. *Mar. Pollut. Bull* 2015, 96, 491–495.

37. Watts AJ, Urbina MA, Corr S, Lewis C, and Galloway TS. Ingestion of Plastic Microfibers by the Crab Carcinus maenas and Its Effect on Food Consumption and Energy Balance. *Environ. Sci. Technol* 2015, 49(24), 14597– 14604.

38. Devriese LI, van der Meulen MD, Maes T, Bekaert K, Paul-Pont I, Frère L, Robbens J, and Vethaak AD, Microplastic contamination in brown shrimp (*Crangon crangon*, Linnaeus 1758) from coastal waters of the Southern North Sea and Channel area. *Mar. Pollut. Bull* 2015, 98(1–2), 179–187.

39. Stasolla G, Innocenti G, and Galil BS. On the diet of the invasive crab Charybdis longicollis Leene, 1938 (Brachyura: Portunidae) in the eastern Mediterranean Sea. *Isr. J. Ecol. Evol* 2016, 61(3–4), 130–134.

40. Watts AJ, Urbina MA, Goodhead R, Moger J, Lewis C, and Galloway TS, Effect of Microplastic on the Gills of the Shore Crab Carcinus maenas. *Environ. Sci. Technol* 2016, 50(10), 5364–5369.

41. Welden NA and Cowie PR. Long-term microplastic retention causes reduced body condition in the langoustine, Nephrops norvegicus. *Environ. Pollut* 2016, 214, 859–865.
42. Hamer J, Gutow L, Köhler A, and Saborowski R. Fate of Microplastics in the Marine Isopod Idotea emarginata. *Environ. Sci. Technol* 2012, 48(22), 13451–13458.
43. Ward JE and Targett NM. Influence of marine microalgal metabolites on the feeding behavior of the blue mussel Mytilus edulis. *Mar. Biol* 1989, 101(3), 313–321.
44. Brilliant MGS and MacDonald BA. Postingestive selection in the sea scallop, Placopecten magellanicus (Gmelin): The role of particle size and density. *J. Exp. Mar. Biol. Ecol* 2000, 253(2), 211–227.
45. Brillant M and MacDonald B. Postingestive selection in the sea scallop (Placopectenmagellanicus) on the basis of chemical properties of particles. *Mar. Biol* 2002, 141(3), 457–465.
46. Ward JE, Levinton JS, and Shumway SE. Particle sorting in bivalves: In vivo determination of the pallial organs of selection. *J. Exp. Mar. Biol. Ecol* 2003, 293(2), 129–149.
47. Browne MA, Dissanayake A, Galloway TS, Lowe DM, and Thompson RC. Ingested Microscopic Plastic Translocates to the Circulatory System of the Mussel, Mytilus edulis (L.). *Environ. Sci. Technol* 2008, 42(13), 5026–5031.
48. Ward JE and Kach DJ. Marine aggregates facilitate ingestion of nanoparticles by suspension-feeding bivalves. *Mar. Environ. Res* 2009, 68(3), 137–142.
49. Von Moos N, Burkhardt-Holm P, and Köhler A. Uptake and Effects of Microplastics on Cells and Tissue of the Blue Mussel Mytilus edulis L. after an Experimental Exposure Environ. *Sci. Technol* 2012, 46(20), 11327–11335.
50. Wegner A, Besseling E, Foekema EM, Kamermans P, and Koelmans AA. Effects of nanopolystyrene on the feeding behavior of the blue mussel (Mytilus edulis L.) *Environ. Toxicol. Chem* 2012, 31, 2490– 2497.
51. Claessens M, Van Cauwenberghe L, Vandegehuchte MB, and Janssen CR. New techniques for the detection of microplastics in sediments and field collected organisms. *Mar. Pollut. Bull* 2013, 70(1–2), 227–233.
52. De Witte B, Devriese L, Bekaert K, Hoffman S, Vandermeersch G, Cooreman K, and Robbens J. Quality assessment of the blue mussel (Mytilus edulis): Comparison between commercial and wild types. *Mar. Pollut. Bull* 2014, 85(1), 146–155.
53. Mathalon A and Hill P. Microplastic fibers in the intertidal ecosystem surrounding Halifax Harbor, Nova Scotia. *Mar. Pollut. Bull* 2014, 81(1), 69–79.
54. Van Cauwenberghe L and Janssen CR. Microplastics in bivalves cultured for human consumption. *Environ. Pollut* 2014, 193, 65–70.
55. Avio CG, Gorbi S, Milan M, Benedetti M, Fattorini D, d'Errico G, Pauletto M, Bargelloni L, and Regoli F. Pollutants bioavailability and toxicological risk from microplastics to marine mussels. *Environ. Pollut* 2015, 198, 211–222.
56. Canesi L, Ciacci C, Bergami E, Monopoli MP, Dawson KA, Papa S, Canonico B, and Corsi I. Evidence for immunomodulation and apoptotic processes induced by cationic polystyrene nanoparticles in the hemocytes of the marine bivalve Mytilus. *Mar. Environ. Res* 2015, 111, 34–40
57. Li J, Yang D, Li L, Jabeen K, and Shi H. Microplastics in commercial bivalves from China. *Environ. Pollut* 2015, 207, 190–195.
58. Rochman CM, Tahir A, Williams SL, Baxa DV, Lam R, Miller JT, Teh FC, Werorilangi S, and Teh SJ. Anthropogenic debris in seafood: Plastic debris and fibers from textiles in fish and bivalves sold for human consumption. *Sci. Rep* 2015, 5, 14340, DOI: 10.1038/srep14340.

59. Sussarellu R, Suquet M, Thomas Y, Lambert C, Fabioux C, Pernet MEJ, Le Goïc N et al. Oyster reproduction is affected by exposure to polystyrene microplastics. *Proc. Natl. Acad. Sci. U. S. A* 2016, 113(9), 2430–2435.

60. Vandermeersch G, Van Cauwenberghe L, Janssen CR, Marques A, Granby K, Fait G, Kotterman MJ et al. A critical view on microplastic quantification in aquatic organisms. *Environ. Res* 2015, 143, 46–55.

61. Davidson K and Dudas SE, Microplastic Ingestion by Wild and Cultured Manila Clams (*Venerupis philippinarum*) from Baynes Sound, British Columbia. *Arch. Environ. Contam. Toxicol* 2016, 71(2), 147–156.

62. Paul-Pont I, Lacroix C, Gonzalez Fernandez C, Hegaret H, Lambert C, Le Goic N, Frere L et al. Exposure of marine mussels *Mytilus* spp. to polystyrene microplastics: Toxicity and influence on fluoranthene bioaccumulation. *Environ. Pollut* 2016, 216, 724–737.

63. Santana MFM, Ascer LG, Custódio MR, Moreira FT, and Turra A. Microplastic contamination in natural mussel beds from a Brazilian urbanized coastal region: Rapid evaluation through bioassessment. *Mar. Pollut. Bull* 2016, 106(1–2), 183–189.

64. Braid HE, Deeds J, DeGreasse SL, Wilson JJ, Osborne J, and Hanner RH. Preying on commercial fisheries and accumulating paralytic shellfish toxins: A dietary analysis of invasive *Dosidicus gigas* (Cephalopoda Ommastrephidae) stranded in Pacific Canada. *Mar. Biol* 2012, 159(1), 25–31.

65. Goldstein MC and Goodwin DS. Gooseneck barnacles (Lepas spp.) ingest microplastic debris in the North Pacific Subtropical Gyre. *PeerJ*, 2014, 1, e841.

66. Carpenter EJ, Anderson SJ, Harvey GR, Miklas HP, and Peck BB. Polystyrene Spherules in Coastal Waters.*Science*, 1972, 178, 749–750.

67. Kartar S, Milne RA, and Sainsbury M. Polystyrene waste in the Severn Estuary. *Science*, 1976, 79(3), 52.

68. Dos Santos J and Jobling M. A model to describe gastric evacuation in cod (*Gadus morhua* L.) fed natural prey. *ICES J. Mar. Sci* 1992, 49(2), 145–154.

69. Boerger CM, Lattin GL, Moore SL, and Moore CJ. Plastic ingestion by planktivorous fishes in the North Pacific Central Gyre. *Mar. Pollut. Bull* 2010, 60(12), 2275–2278.

70. Davison P, Asch RG, Davison P, and Asch RG. Plastic ingestion by mesopelagic fishes in the North Pacific Subtropical Gyre. *Mar. Ecol.: Prog. Ser* 2011, 432, 173–180.

71. Possatto FE, Barletta M, Costa MF, do Sul JAI, and Dantas DA. Plastic debris ingestion by marine catfish: An unexpected fisheries impact. *Mar. Pollut. Bull* 2011, 62(5), 1098–1102.

72. Dantas DV, Barletta M, da Costa MF, The seasonal and spatial patterns of ingestion of polyfilament nylon fragments by estuarine drums (Sciaenidae). *Environ. Sci. Pollut. Res* 2012, 19(2), 600–606.

73. Ramos JAA, Barletta M, and Costa MF. Movement patterns of catfishes (Ariidae) in a tropical semi-arid estuary. Aquat. *Biol* 2012, 17, 29–34.

74. Choy CA and Drazen JC. Plastic for dinner? Observations of frequent debris ingestion by pelagic predatory fishes from the central North Pacific. *Mar. Ecol.: Prog. Ser* 2013, 485, 155–163.

75. Foekema EM, De Gruijter C, Mergia MT, van Franeker JA, Murk AJ, and Koelmans AA. Inventory of the presence of plastics in the digestive tract of North Sea fishes Environ. *Sci. Technol* 2013, 47(15), 8818–8824.

76. Gassel M, Harwani S, Park J-S, and Jahn A. Detection of nonylphenol and persistent organic pollutants in fish from the North Pacific Central Gyre. *Mar. Pollut. Bull* 2013, 73, 231–242.

77. Lusher AL, McHugh M, and Thompson RC. Occurrence of microplastics in the gastrointestinal tract of pelagic and demersal fish from the English Channel; *Mar. Pollut. Bull* 2013, 67(1–2), 94–99.

78. Saji Kumar KK, Ragesh N, Remya R, and Mohamed KS. Scope for mechanized fishing of teleosts with light attraction in Southeastern Arabian Sea; Marine Fisheries Information Service; Technical and Extension Series 2013, 217, 13.

79. Kripa V, Nair PG, Dhanya AM, Pravita VP, Abhilash S, Mohamed AA, Vijayan D et al. Microplastics in the gut of anchovies caught from the mud bank area of Alappuzha, Kerala; Marine Fisheries Information Service; Technical and Extension Series, 2014, 219, 27–28.

80. Sulochanan B, Bhat GS, Lavanya S, Dineshbabu AP, and Kaladharan P. A preliminary assessment of ecosystem process and marine litter in the beaches of Mangalore. *Indian J. Geo-Mar. Sci* 2013, 43(9), 1–6.

81. Avio CG, Gorbi S, and Regoli F. Experimental development of a new protocol for extraction and characterization of microplastics in fish tissues: First observations in commercial species from Adriatic Sea. *Mar. Environ. Res* 2015,111, 18–26.

82. Collard F, Gilbert B, Eppe G, Parmentier E, and Das K. Detection of Anthropogenic Particles in Fish Stomachs: An Isolation Method Adapted to Identification by Raman Spectroscopy. *Arch. Environ. Contam. Toxicol* 2015, 69(3), 331–339.

83. Neves D, Sobral P, Ferreira JL, and Pereira T. Ingestion of microplastics by commercial fish off the Portuguese coast. *Mar. Pollut. Bull* 2015, 101, 119–126.

84. Phillips MB and Bonner TH. Occurrence and amount of microplastic ingested by fishes in watersheds of the Gulf of Mexico. *Pollut. Bull* 2015, 100, 264–269.

85. Romeo T, Pietro B, Peda C, Consoli P, Andaloro F, and Fossi MC. First evidence of presence of plastic debris in stomach of large pelagic fish in the Mediterranean Sea. *Mar. Pollut. Bull* 2015, 95, 358–361.

86. Bellas J, Martínez-Armental J, Martínez-Cámara A, Besada V, and Martínez-Gómez C. Ingestion of microplastics by demersal fish from the Spanish Atlantic and Mediterranean coasts. *Mar. Pollut. Bull* 2016, 109(1), 55–60.

87. Cannon SME, Lavers JL, and Figueiredo B. Plastic ingestion by fish in the Southern Hemisphere: A baseline study and review of methods. *Mar. Pollut. Bull* 2016, 107, 286–291.

88. Miranda DDA and de Carvalho-Souza GF. Are we eating plastic-ingesting fish? *Mar. Pollut. Bull* 2016, 103, 109–114.

89. Lusher AL, O'Donnell C, Officer R, and O'Connor I. Microplastic interactions with North Atlantic mesopelagic fish. *ICES J. Mar. Sci* 2016, 73(4), 1214–1225.

90. Nadal MA, Alomar C, and Deudero S. High levels of microplastic ingestion by the semipelagic fish bogue Boops boops (L.) around the Balearic Islands. *Environ. Pollut* 2016, 214, 517–523.

91. Naidoo T, Smit AJ, and Glassom D. Plastic ingestion by estuarine mullet Mugil cephalus (Mugilidae) in an urban harbour, KwaZulu-Natal, *South Africa. Afr. J. Mar. Sci* 2016, 38(1), 145–149.

92. Pedà C, Caccamo L, Fossi MC, Gai F, Andaloro F, Genovese L, Perdichizzi A, Romeo T, and Maricchiolo G. Intestinal alterations in European sea bass Dicentrarchus labrax (Linnaeus, 1758) exposed to microplastics: Preliminary results. *Environ. Pollut* 2016, 212, 251–256.

93. Rummel CD, Löder MG, Fricke NF, Lang T, Griebeler EM, Janke M, and Gerdts G. Plastic ingestion by pelagic and demersal fish from the North Sea and Baltic Sea. *Mar. Pollut. Bull* 2016, 102, 134–141.

94. Ferreira P, Fonte E, Soares ME, Carvalho F, and Guilhermino L. Effects of multi-stressors on juveniles of the marine fish Pomatoschistus microps: Gold nanoparticles, microplastics and temperature. *Aquat. Toxicol* 2016, 170, 89–103.

95. Tourinho PS, do Sul JAI, and Fillmann G. Is marine debris ingestion still a problem for the coastal marine biota of southern Brazil..*Mar. Pollut. Bull* 2010, 60(3), 396–401.

96. Van Franeker JA, Blaize C, Danielsen J, Fairclough K, Gollan J, Guse N, Hansen PL et al. Monitoring plastic ingestion by the northern fulmar Fulmarus glacialis in the North Sea. *Environ. Pollut* 2011, 159(10), 2609–2615.

97. Besseling E, Foekema EM, van Franeker JA, Leopold MF, Kuhn S, Rebolledo EB, Heße E et al. Microplastic in a macro filter feeder: Humpback whale Megaptera novaeangliae. *Mar. Pollut. Bull* 2015, 95(1), 248–252.

98. Lusher AL, Herńandez-Milían G, O'Brien J, Berrow S, O'Connor I, and Officer R. Microplastic and macroplastic ingestion by a deep diving, oceanic cetacean: The True's beaked whale *Mesoplodon mirus*. *Environ. Pollut* 2015, 199, 185–191.

99. Rosenkranz P, Chaudhry Q, Stone V, and Fernandes TF. A comparison of nanoparticle and fine particle uptake by Daphnia magna. *Environ. Toxicol. Chem* 2009, 28(10), 2142–2149.

100. Imhof HK, Schmid J, Niessner R, Ivleva NP, and Laforsch C. Contamination of beach sediments of a subalpine lake with microplastic particles. *Limnol. Oceanogr.: Methods*, 2012, 10(7), 524– 537.

101. Sanchez W, Bender C, and Porcher JM. Wild gudgeons (*Gobio gobio*) from French rivers are contaminated by microplastics: Preliminary study and first evidence. *Environ. Res* 2014, 128, 98–100.

102. Faure F, Demars C, Wieser O, Kunz M, and De Alencastro LF. Plastic pollution in Swiss surface waters: Nature and concentrations, interaction with pollutants. *Environ. Chem* 2015, 12(5), 582–591.

103. Biginagwa FJ, Mayoma BS, Shashoua Y, Syberg K, and Khan FR. First evidence of microplastics in the African Great Lakes: Recovery from Lake Victoria Nile perch and Nile tilapia. *J. Great Lakes Res* 2016, 42(1), 146–149.

104. Lonnstedt OM and Eklov P. Environmentally relevant concentrations of microplastic particles influence larval fish ecology. *Science*, 2016, 352(6290), 1213–1216.

105. Lu Y, Zhang Y, Deng Y, Jiang W, Zhao Y, Geng J, Ding L, and Ren H. Uptake and Accumulation of Polystyrene Microplastics in Zebrafish (Danio rerio) and Toxic Effects in Liver. *Environ. Sci. Technol* 2016, 50(7), 4054–4060.

106. Wardrop P, Shimeta J, Nugegoda D, Morrison PD, Miranda A, Tang M, and Clarke BO. Chemical Pollutants Sorbed to Ingested Microbeads from Personal Care Products Accumulate in Fish. *Environ. Sci. Technol* 2016, 50(7), 4037–4044.

107. Peters CA and Bratton SP. Urbanization is a major influence on microplastic ingestion by sunfish in the Brazos River Basin, Central Texas, USA. *Environ. Pollut* 2016, 210, 380–387.

108. Besseling E, Wang B, Lurling M, and Koelmans AA. Nanoplastic Affects Growth of S. obliquus and Reproduction of D. magna. *Environ. Sci. Technol* 2014, 48(20), 12336–12343.

109. Ogonowski M, Schur C, Jarsen A, and Gorokhova E. The Effects of Natural and Anthropogenic Microparticles on Individual Fitness in Daphnia magna.*PLoS One* 2016, 11(5), e0155063.

110. Nasser F and Lynch I. Secreted protein eco-corona mediates uptake and impacts of polystyrene nanoparticles on Daphnia magna. *J. Proteomics*, 2016, 137, 45–51.

111. Rehse S, Kloas W, and Zarfl C. Short-term exposure with high concentrations of pristine microplastic particles leads to immobilisation of *Daphnia magna*. *Chemosphere* 2016, 153, 91–99.

112. Huerta Lwanga E, Gertsen H, Gooren H, Peters P, Salanki T, van der Ploeg M, Besseling E, Koelmans AA, and Geissen V. Potential risk of microplastics transportation into ground water. *Environ. Sci. Technol* 2016, 50(5), 2685–2691.

113. Cedervall T, Hansson LA, Lard M, Frohm B, and Linse S. Food chain transport of nanoparticles affects behaviour and fat metabolism in fish. *PLoS One*, 2012, 7(2), e32254.

114. Farrell P and Nelson K. Trophic level transfer of microplastic: *Mytilus edulis* (L.) to *Carcinus maenas* (L.). *Environ. Pollut* 2013, 177, 1–3.

115. Batel A, Linti F, Scherer M, Erdinger L, and Braunbeck T. Transfer of benzo[a]pyrene from microplastics to Artemia nauplii and further to zebrafish via a trophic food web experiment: CYP1A induction and visual tracking of persistent organic pollutants. *Environ. Toxicol. Chem* 2016, 35(7), 1656–1666.

116. Bhattacharya P, Lin S, Turner JP, and Ke P-C. Physical adsorption of charged plastic nanoparticles affects Algal Photosynthesis. *J. Phys. Chem. C*, 2010, 114(39), 16556–16561.

117. Long M, Moriceau B, Gallinari M, Lambert C, Huvet A, Raffray J, and Soudant P. Interactions between microplastics and phytoplankton aggregates: Impact on their respective fates. *Mar. Chem* 2015, 175, 39–46.

118. Lagarde F, Olivier O, Zanella M, Daniel P, Hiard S, and Caruso A. Microplastic interactions with freshwater microalgae: Hetero-aggregation and changes in plastic density appear strongly dependent on polymer type. *Environ. Pollut* 2016, 215, 331–339.

119. Gutow L, Eckerlebe A, Gimenez L, and Saborowski R. Microplastic interactions with freshwater microalgae: Hetero-aggregation and changes in plastic density appear strongly dependent on polymer type. *Environ. Sci. Technol* 2016, 50, 915–923.

120. Bakir A, Rowland SJ, and Thompson RC. Enhanced desorption of persistent organic pollutants from microplastics under simulated physiological conditions. *Environ. Pollut* 2014, 185, 16–23.

121. Rochman CM. In: Bergmann M, Gutow L and Klages M (eds) *Marine Anthropogenic Litter*. Springer, Berlin. 2015, pp. 117–140.

122. Rochman CM, Ecologically relevant data are policy-relevant data Science, 2016, 352(6290), 1172.

123. Hidalgo-Ruz V, Gutow L, Thompson RC, and Thiel M. Microplastics in the Marine Environment: A Review of the Methods Used for Identification and Quantification. *Environ.Sci. Technol* 2012, 46(6), 3060–3075.

124. Sæther BS and Jobling M. Gastrointestinal evacuation of inert particles by turbot, Psetta maxima: Evaluation of the X-radiographic method for use in feed intake studies. *Aquat. Living Resour* 1997, 10(6), 359–364.

125. McGaw IJ and Curtis DL. A review of gastric processing in decapod crustaceans. *J. Comp. Physiol, B*, 2013, 183(4), 443–465.

126. Powell MD and Berry AJ. Ingestion and regurgitation of living and inert materials by the estuarine copepod *Eurytemora affinis* (Poppe) and the influence of salinity. Estuarine, *Coastal Shelf Sci* 1990, 31(6), 763–773.

127. Bromley PJ, The role of gastric evacuation experiments in quantifying the feeding rates of predatory fish. *Rev. Fish Biol. Fish* 1994, 4(1), 36–66.

128. Bowman RE. Effect of regurgitation on stomach content data of marine fishes Environ. *Biol. Fishes*, 1986, 16(1), 171–181.

129. Bowen SH. In: Murphy BR and Willis DW (eds) *Fisheries Techniques*, 2nd edn. American Fisheries Society, Bethesda, MD. 1996, pp. 513–532.

130. Daan N. A quantitative analysis of the food intake of North Sea cod, Gadus Morhua. *Neth. J. Sea Res* 1973, 6(4), 479–517.

131. Barltrop D and Meek F. Effect of particle size on lead absorption from the Gut. *Arch. Environ. Health*, 1979, 34(4), 280–385.

132. Reineke JJ, Cho DY, Dingle Y-T, Morello AP, Jacob J, Thanos CG, and Mathiowitz E. Unique insights into the intestinal absorption, transit, and subsequent biodistribution of polymer-derived microspheres. *Proc. Natl. Acad. Sci. U. S. A* 2013, 110(34), 13803–13808.

133. Andrady AL. In: Bergmann M, Gutow L and Klages M (eds) *Marine Anthropogenic Litter.* Springer, Berlin. 2015, pp. 57–72.

134. Nakashima S, Sturgeon RE, Willie SN, and Berman SS. Acid digestion of marine samples for trace element analysis using microwave heating. *Analyst*, 1988, 113, 159–163.

135. Dehaut A, Cassone A-L, Frere L, Hermabessiere L, Himber C, Rinnert E, Riviere G et al. Microplastics in seafood: Benchmark protocol for their extraction and characterization. *Environ. Pollut* 2016, 215, 223–233.

136. Jin Y, Li H, Mahar RB, Wang Z, and Nie Y. Effects and model of alkaline waste activated sludge treatment. *J. Environ. Sci* 2009, 21, 279–284.

137. Nuelle MT, Dekiff JH, Remy D, and Fries E. A new analytical approach for monitoring microplastics in marine sediments. *Environ. Pollut* 2014, 184, 161–169.

138. Stojicic S, Zivkovic S, Qian W, Zhang H, and Haapasalo M. Tissue Dissolution by Sodium Hypochlorite: Effect of Concentration, Temperature, Agitation, and Surfactant. *J. Endod* 2010, 36(9), 1558–1562.

139. Duncan E. Presented in part at Micro 2016, Lanzarote, Canary Islands, Spain, May, 2016.

140. Catarino AI, Thompson R, Sanderson W, and Henry TB. Development and optimization of a standard method for extraction of microplastics in mussels by enzyme digestion of soft tissues. *Environ. Toxicol. Chem* 2017, 36 (4), 947–951.

141. Scholz-Bottcher B. Presented in part at Micro 2016, Lanzarote, Canary Islands, Spain, May, 2016.

142. Horton AA, Svendsen C, Williams RJ, Spurgeon DJ, and Lahive E. Large microplastic particles in sediments of tributaries of the River Thames, UK – Abundance, sources and methods for effective quantification. *Mar. Pollut. Bull* 2017, 114(1), 218–226.

143. Gilftllan LR, Ohman MD, Doyle MJ, and Watson W. California Cooperative Oceanic and Fisheries Investigations, 2009, 50, 123–133.

144. Sgier L, Freimann R, Zupanic A, and Kroll A. Flow cytometry combined with viSNE for the analysis of microbial biofilms and detection of microplastics Nat. *Commun* 2016, 7, 11587, DOI: 10.1038/ncomms11587.

145. Reisser J, Shaw J, Hallegraeff G, Proietti M, Barnes DA, Thums M, Wilcox C, Hardesty BD, and Pattiaratchi C. Millimeter-Sized Marine Plastics: A New Pelagic Habitat for Microorganisms and Invertebrates. *PLoS ONE 9*(6): e100289. doi:10.1371/journal.pone.0100289

146. Löder M and Gerdts G. In: Bergmann M, Gutow L, Klages M (eds) *Marine Anthropogenic Litter.* Springer, Berlin, 2015, pp. 201–227.

147. Remy F, Collard F, Gilbert B, Compère P. Eppe G, and Lepoint G. When Microplastic Is Not Plastic: The Ingestion of Artificial Cellulose Fibers by Macrofauna Living in Seagrass Macrophytodetritus. *Environ. Sci. Technol* 2015, 49 (18), 11158–11166.

148. Rocha-Santos T and Duarte AC. A critical overview of the analytical approaches to the occurrence, the fate and the behavior of microplastics in the environment. *TrAC, Trends Anal. Chem* 2015, 65, 47–53.

149. Song YK, Hong SH, Jang M, Han GM, Rani M, Lee J, and Shim WJ. A comparison of microscopic and spectroscopic identification methods for analysis of microplastics in environmental samples. *Mar. Pollut. Bull* 2015, 93(1–2), 202–209.

150. Eriksen M, Mason S, Wilson S, Box C, Zellers A, Edwards W, Farley H, and Amato S. Microplastic pollution in the surface waters of the Laurentian Great Lakes. *Mar. Pollut. Bull* 2013, 77(1–2), 177–182.

151. Lima ARA, Costa MF, and Barletta M. Distribution patterns of microplastics within the plankton of a tropical estuary. *Environ. Res* 2014, 132, 146–155.

152. Lima ARA, Barletta M, and Costa MF. Seasonal-Dial Shifts of Ichthyoplankton Assemblages and Plastic Debris around an Equatorial Atlantic Archipelago Front. *Environ.Sci* 2016, 4, 56.

153. Hamza AA, Sokkar TZN, and Ramadan WA. On the microinterferometric determination of refractive indices and birefringence of fibres. *Pure Appl. Opt* 1992, 1(6), 321

154. Fries E, Dekiff JH, Willmeyer J, Nuelle MT, Ebert M, and Remy D. Identification of polymer types and additives in marine microplastic particles using pyrolysis-GC/MS and scanning electron microscopy. *Environ. Sci.: Processes Impacts*, 2013, 15(10), 1949–1956.

155. Dumichen E, Barthel A-K, Braun U, Bannick CG, Brand K, Jekel M, and Senz R. Analysis of polyethylene microplastics in environmental samples, using a thermal decomposition method. *Water Res* 2015, 85, 451–457.

156. Hintersteiner I, Himmelsbach M, and Buchberger WW. Characterization and quantitation of polyolefin microplastics in personal-care products using high-temperature gel-permeation chromatography. *Anal. Bioanal. Chem* 2015, 407, 1253–1259.

157. Fischer M. Presented in part at Micro 2016, Lanzarote, Canary Islands, Spain, May, 2016.

158. Browne MA, Crump P, Niven SJ, Teuten E, Tonkin A, Galloway T, and Thompson RC. Accumulation of Microplastic on Shorelines Woldwide: Sources and Sinks. *Environ. Sci. Technol* 2011, 45(21), 9175–9179.

159. Dris R, Gasperi J, Saad M, Mirande C, and Tassin B. Synthetic fibers in atmospheric fallout: A source of microplastics in the environment?. *Mar. Pollut. Bull* 2016, 104, 290–293.

160. Woodall LC, Gwinnett C, Packera M, Thompson RC, Robinson LF, and Paterson GLJ. Using a forensic science approach to minimize environmental contamination and to identify microfibres in marine sediments. *Mar. Pollut. Bull* 2015, 95(1), 40–46.

161. de Śa LC, Lúis LG, and Guilhermino L. Effects of microplastics on juveniles of the common goby (*Pomatoschistus microps*): Confusion with prey, reduction of the predatory performance and efficiency, and possible influence of developmental conditions. *Environ. Pollut* 2015, 196, 359–362.

162. Rochman CM, Hoh E, Kurobe T, and Teh SJ. Ingested plastic transfers hazardous chemicals to fish and induces hepatic stress. *Sci. Rep* 2013, 3, 03263.

163. Rochman CM, Kurobe T, Flores I, and Teh SJ. Early warning signs of endocrine disruption in adult fish from the ingestion of polyethylene with and without sorbed chemical pollutants from the marine environment. *Sci. Total Environ* 2014, 493, 656–661.

164. Bergami E, Bocci E, Vannuccini ML, Monopoli M, Salvati A, Dawson KA, and Corsi I. Nano-sized polystyrene affects feeding, behavior and physiology of brine shrimp *Artemia franciscana larvae*. *Ecotoxicol. Environ. Saf* 2016, 123, 18–25.

165. R Core Team, R: A language and environment for statistical computing, R Foundation for Statistical Computing, 2013, Vienna, Austria, http://www.R-project.org/.

166. EFSA CONTAM Panel EFSA Panel on Contaminants in the food chain. *EFSA J* 2016, 14(6), e04501, DOI: 10.2903/j.efsa.2016.4501.

167. Urbina MA, Watts AJR, and Reardon EE. Labs should cut plastic waste too. *Nature* 2015, 528, 479, DOI: 10.1038/528479c.

168. Mattsson K, Ekvall MT, Hansson LA, Linse S, Malmendal A, and Cedervall T. Altered Behavior, Physiology, and Metabolism in Fish Exposed to Polystyrene Nanoparticles. *Environ. Sci. Technol* 2015, 49(1), 553–561.

169. Rummel CD, Adolfsson-Erici M, Jahnke A, and MacLeod L. No measurable "cleaning" of polychlorinated biphenyls from Rainbow Trout in a 9 week depuration study with dietary exposure to 40% polyethylene microspheres. *Environ. Sci.: Processes Impacts* 2016, 18, 788–795.

170. Welden NA and Cowie P. Long-term microplastic retention causes reduced body condition in the langoustine, *Nephrops norvegicus. Environ. Pollut* 2016, 218, 895–900.

171. Davarpanah E and Guilhermino L. Single and combined effects of microplastics and copper on the population growth of the marine microalgae *Tetraselmis chuii. Estuarine, Coastal Shelf Sci* 2015, 167, 269–275.

172. Mazurais D, Ernande B, Quazuguel P, Severe A, Huelvan C, Madec L, Mouchel O et al. Evaluation of the impact of polyethylene microbeads ingestion in European sea bass (*Dicentrarchus labrax*) larvae. *Mar. Environ. Res* 2015, 112, 78–85.

173. Lúis LG, Ferreira P, Fonte E, Oliveira M, and Guilhermino L. Does the presence of microplastics influence the acute toxicity of chromium(VI) to early juveniles of the common goby (*Pomatoschistus microps*)? A study with juveniles from two wild estuarine populations.Aquat. *Toxicol* 2015, 164, 163–174.

174. Oliveria M, Riberio A, Hylland K, and Guilhermino L. Single and combined effects of microplastics and pyrene on juveniles (0+ group) of the common goby *Pomatoschistus microps* (Teleostei, *Gobiidae). Ecol. Indic*, 2013, 34, 641–647.

175. Clarke JR, Cole M, Lindeque PK, Fileman E, Blackford J, Lewis C, Lenton TM, and Galloway TS. Marine microplastic debris: A targeted plan for understanding and quantifying interactions with marine life. *Front. Ecol. Environ* 2016, 14, 317–324.

SECTION V

Microplastics and POPs

CHAPTER 9

Persistent Organic Pollutants (POPs)

A Real Threat to the Environment

M. Humam Zaim Faruqi and Faisal Zia Siddiqui

CONTENTS

9.1 INTRODUCTION

The contamination of world's aquatic environment is an area of growing concern and research. Contaminants of emerging concern in the global environment comprise more than 40,000 different chemicals, which is increasing by six additional chemicals on a daily basis [1,2]. Moreover, plastic materials are estimated to release about 35–917 tonnes of chemical additives into the global marine environment each year [3].

Degradation of plastic products into smaller fragments such as microplastics or nanoplastics and their subsequent passage into the marine ecosystems result in their transport even to remote locations on Earth's surface. Plastics are known to concentrate and transport toxic chemicals that are either added during their production as additives or are sorbed into the plastic matrix while microplastic fragments come in contact with polluted water.

Certain chemicals in Earth's ecosystem are resistant to environmental degradation in terms of chemical, biological and photolytic reactions [4]. Such chemicals in our environment that exist for a longer time duration are called persistent organic pollutants (POPs). POPs are persistent in the environment, having long half-lives in soils, sediments or the atmosphere. There is no universal consensus on how long the half-life in a given medium should be for the term "persistent" to be used. However, a POP could have a half-life of years or even decades in soil or sediment and several days in the atmosphere [5].

POPs are priority pollutants consisting of pesticides (dichlorodiphenyltrichloroethane [DDT] and its metabolites, toxaphene, chlordane), industrial chemicals (polychlorinated biphenyls [PCBs], polybrominated diphenyl ethers [PBDEs]), byproducts of industrial processes (dioxins and furans) and precursors to important plastics (bisphenol A) [4].

POPs are typically hydrophobic and lipophilic in nature. Plastic debris acts as a vector for organic contaminants owing to their hydrophobic nature. POPs partition strongly to solids in aquatic systems and soils, avoiding the aqueous phase. In organisms, these partition into lipids rather than entering the aqueous medium of cells and become stored in fatty tissue. POPs may also volatilize from soils, vegetation and water bodies into the atmosphere and travel long distances before being deposited because of their resistance to breakdown reactions in air. POPs can partition between particles and aerosol depending upon ambient temperature and physicochemical properties of the chemical. The unique combination of resistance to metabolism and lipophilicity leads to the accumulation of POPs in food chains [5].

9.2 SOURCES AND PERSISTENCE OF POPs

Because of the long-range transport and harmful effects on human, flora and fauna, several global efforts have been made to reduce future environmental impacts of POPs. These include the 1998 Aarhus Protocol on POPs and the 2001 Stockholm Convention on POPs to eliminate or control their releases. Prior knowledge about sources and emission rates of POPs into the environment is essential if environmental burdens are to be reduced and quantitative source-receptor relationships at regional and global scales are to be developed. Most POPs can be broadly classified according to the source of generation as *intentionally produced* or *accidentally produced* [5]. Intentionally produced POPs may be subdivided into several subgroups such as pesticides and industrial chemicals such as chlordane, DDT, PCBs, PBDEs and others. Accidentally produced POPs are usually separated between combustion and chemical-industrial processes. Some examples are polychlorinated dibenzo-*p*-dioxins (PCDD), polychlorinated dibenzofurans (PCDFs) and polycyclic aromatic hydrocarbons (PAHs) [5].

Hexachlorobenzene (HCB) and PCBs are examples of POPs that are used as industrial chemicals but are also inadvertently formed as by-products in combustion and other processes [6–8]. The Stockholm Convention on POPs lists HCB and PCB not only as intentionally produced POPs but also as unwanted by-products and therefore requires identification and quantification of their sources and establishment of release inventories from non-intentional production [5].

POPs may also be classified into the following categories according to the source of its generation [4]:

- Chemicals that are subject to elimination of production and use (e.g., Aldrin, hexachlorobenzene, chlordane, HCB, PCB)
- Chemicals that are restricted in production and use (e.g., DDT, perfluorooctanesulfonic acid)
- Chemicals that are unintentionally produced (e.g., pentachlorobenzene, HCB, PCB)
- Chemicals under investigation for persistence (e.g., PAHs, chlorinated naphthalenes, hexachlorobutadiene)

In addition to forestry, agriculture, horticulture, municipal, industrial and medical activities, natural sources such as volcanic activities and vegetation fires also contribute these chemicals to the ecosystem. Further, water and direct contact sources are also responsible for POPs contamination [4,9].

The lipophilic property of POPs is responsible for long-time persistence, transport from one organism to another and bioaccumulation at higher concentrations. Some POPs, such as perfluorooctanesulfonic acid (PFOS), are water soluble, which has caused their detection in municipal wastewater and drinking water samples [4]. Halogenated POPs are more resistant to degradation reactions, have the ability to associate with aerosols and, hence, transport across long distances. The fate of transportation depends on meteorological conditions, physicochemical properties, and the removal process by photochemically driven reactions [10]. The transfer of POPs from Earth's surface to the atmosphere takes place in two steps, that is, transformation from liquid or solid state to vapors and dispersion by mixing [11,12]. Thus, POPs can reach at far distances from their sources to even remote locations such as Antarctica and the Arctic Circle.

9.3 CURRENT SCENARIO OF POPs

Exposure to POPs is known to cause serious health problems such as obesity, hormonal disruption, cancer, diabetes, cardiovascular diseases, and reproductive and neurological ailments, in addition to developing defects in women embryo [4]. As per the World Health Organization (WHO), 2,3,7,8-tetrachlorodibenzo-p-dioxin (2,3,7,8-TCDD) is the most toxic POP [4,13]. The US Environmental Protection Agency (EPA) describes dioxins as a major cause of cancer [4]. Some POPs which have been recognized as of increasing concern worldwide are described here.

9.3.1 Polychlorinated Biphenyls (PCBs)

Polychlorinated biphenyls (PCBs) are commonly considered as the key representative of "industrial" POPs and have been produced in high volumes by the chemical and process industries [14]. The high utility of PCBs is on account of their chemical stability. PCBs are used as coolants and insulators in transformers and capacitors, as plasticizers in paints and cements and as stabilizing additives in flexible polyvinyl chloride (PVC) coatings. Industrial POPs such as polybrominated diphenyl ethers (PBDEs) share some of the characteristics as PCBs, with respect to their sources and emissions. Although the production of PCBs peaked around the 1970s, the current annual production trends of PBDEs rival the historical peak of PCBs [15]. PCBs are known to cause disturbances in thyroid hormone, sex steroid hormone and cortisol, resulting in behavioral and morphological changes [4,16].

9.3.2 Bisphenol A (BPA)

Bisphenol A (BPA) is a precursor to certain plastics such as polycarbonates and epoxy resins. BPA is used to protectively coat metal cans to prevent contamination and extend product shelf life. It is also added to many common products such as plastic eating utensils, toys, eyeglasses and office products. BPA is released into the environment, especially surface water, when these products go to landfills. It can be ingested or even absorbed through skin contact. BPA has been found to contaminate water supplies, dust and air, with the primary route of exposure being the leaching of BPA into food from incomplete polymerization of epoxy resins and polycarbonate plastic or the degradation of weak ester bonds that link BPA monomers [17].

9.3.3 Polycyclic Aromatic Hydrocarbons (PAHs)

Polycyclic aromatic hydrocarbons (PAHs) are non-polar, lipophilic organic compounds that are generally insoluble in water, thus limiting their mobility in the environment. PAHs are abundant around the globe. Two- or four-ringed PAHs volatilize to appear in the atmosphere in gaseous form [18]. The predominant exposure of humans to PAHs occurs from burning solid fuels such as coal and biofuels for domestic requirements such as cooking and heating [19]. Moreover, tobacco smoke is known to contribute about 90% of indoor PAH levels. The emissions from the transport sector can be a substantial outdoor source of PAHs in contributing to particulate air pollution [20]. The concentration of PAHs in rivers and marine bodies may depend on various factors such as proximity to municipal and industrial discharge points, wind direction and distance from major urban roadways [21]. Adverse effects of PAHs on human health include cancer, cardiovascular disorders, poor fetal growth, reduced immunity and poor neurological development.

9.4 HEALTH AND ENVIRONMENTAL EFFECTS OF POPs

Several studies have been conducted on the effect of POPs on human health. Inhalation and ingestion of dust and air has been described as the major source of POPs in humans. An increase in hypertension was reported due to higher concentration of PCBs and organochlorine (OC) POPs in Arctic populations [22]. Langer et al. described diabetic problems in children because of pesticides and PCBs [23].

Earth's environment is affected by POPs through biotic, abiotic, social or technological interferences and subsequent degradation of ecological balance. The environmental problems and disturbances in aquatic ecosystems observed in the North Sea, Baltic Sea, Great Lakes and the Arctic Sea during the 1980s and 1990s were attributed to higher POP concentrations in these areas [4]. The effects of POP concentration in living species include shift in sex ratios, impaired fertility, wildlife cancer and various other physical abnormalities. POPs have been found to affect biochemical processes (immune and endocrine systems, reproduction and development) and also impact tissue residue levels in several top trophic level species such as seabirds, polar bears, polar fox and sled dogs [4,24]. Climatic changes have also been observed in food webs, ice and snow melt, lipid dynamics and organic carbon cycling.

One of the important phenomena described by POPs is the "grasshopper effect" involving cyclic volatilization and condensation. These chemicals tend to migrate from warmer climates to colder regions and settle down in extremely cool situations, but evaporate when the temperature rises. The presence of this cyclic migration of POPs contributes to global warming.

9.5 FUTURE PROSPECTS

Some key properties of POPs control their fate in the environment, and environmental chemists can make reasonable predictions about their fate and behavior based on these. Such properties include aqueous solubility; vapor pressure; partition coefficients between water–solid, air–solid and air–liquid; and half-lives in air, soil and water [5]. However, there are often wide variations in these properties, leading to uncertainties in the precise behavior of these POPs.

The human population as well as industrial and agricultural activities are increasing continuously, which add POPs to the environment. The regulations to control POPs remain limited to reports, research papers and books, especially in developing countries [4]. However, several research studies have attempted to develop models to predict future perspectives of POPs, depending upon their physical, chemical and mechanistic behavior. Application of various modeling tools have been made for assessment of climate change, assessment of risk profiles of POPs, evaluation of ecological risks in populations and aquatic communities and also review of future challenges posed by POPs based on past understandings and increased leaching of POPs from landfills [4,25–27].

The immediate requirement is to develop remediation and control methods for POPs before their menace further worsens. Some important ways to control the increasing POPs concentration across the environment are described as follows [4]:

- Development of remediation measures by altering the genetic structures of microbes in order to make them capable for bioremediation of POPs.
- Reduction in the production and use of POPs by application of stringent bans on more harmful chemicals and phasing out usage of others.
- Development of effective techniques such as adsorption, solvent extraction, alkali metal reduction, incineration, solidification and pyrolysis for removal of POPs.
- Additionally, more advanced experimental conditions should be developed in electrochemical and membrane technologies to address POPs removal from water resources.

REFERENCES

1. Crawford CB, and Quinn B. *Microplastic Pollutants* n.p.: Elsevier Inc, 2017.
2. Halden RU. Epistemology of contaminants of emerging concern and literature meta-analysis. *J Hazard Mater.* 2015, 282, 2–9.
3. Suhrhoff TJ, and Scholz-Bottcher BM. Qualitative impact of salinity, UV radiation, turbulence on leaching of organic plastic additives from four common plastics – A lab experiment. *Mar Pollut Bull.* 2016, 102, 84–94.
4. Alharbi OML, Basheer AA, Khattab RA, Ali I. Health and environmental effects of persistent organic pollutants. *J Mol Liq.* 2018, 263, 442–453.
5. Jones KC, and de Voogt P. Persistent organic pollutants (POPs): State of the science. *Environ Pollut.* 1999, 100, 209–221.
6. Bailey RE. Global hexachlorobenzene emissions. *Chemosphere* 2001, 43, 167–182.
7. Brown JF Jr., Frame GM, Olson DR, and Webb JL. The sources of the coplanar PCBs. *Organohalogen Compd.* 1995, 26, 427–430.
8. Lohmann R, Northcott GL, and Jones KC. Assessing the contribution of diffuse domestic burning as a source of PCDD/Fs, PCBs, and PAHs to the UK atmosphere. *Environ Sci Technol.* 2000, 34(14), 2892–2899.
9. El-Shahawi MS, Hamza A, Bashammakh AS, and Al-Saggaf WT. An overview on the accumulation, distribution, transformations, toxicity and analytical methods for the monitoring of persistent organic pollutants. *Talanta* 2010, 80, 1587–1597.

10. Foreman WT, Majewski MS, Goolsby DA, Wiebe FW, and Coupe RH. Pesticides in the atmosphere of the Mississippi River Valley, part II – air. *Sci Total Environ.* 2000, 248, 213–226.

11. Beyer A, Mackay D, Matthies M, Wania F, and Webster E. Assessing long-range transport potential of persistent organic pollutants. *Environ Sci Technol.* 2000, 34(4), 699–703.

12. Gavrilescu M. Fate of pesticides in the environment and its bioremediation. *Eng Life Sci.* 2005, 5(6), 497–526.

13. Jain CK, and Ali I. Determination of pesticides in soils, sediment and water systems by gas chromatography. *Int J Environ Anal Chem.* 1997, 68, 83–101.

14. Breivik K, Alcock R, Li Y-F, Bailey RE, Fiedler H, Pacyna JM. Primary sources of selected POPs: Regional and global scale emission inventories. *Environ Pollut.* 2004, 128, 3–16.

15. Breivik K, Sweetman A, Pacyna JM, and Jones KC. Towards a global historical emission inventory for selected PCB congeners – a mass balance approach: 1 Global production and consumption. *Sci Total Environ.* 2002, 290, 181–198.

16. Jenssen BM. Endocrine-disrupting chemicals and climate change: A worst-case combination for Arctic marine mammals and seabirds? *Environ Health Perspect.* 2006, 114, 76–80.

17. Metz CM. Bisphenol A: Understanding the controversy. *Workplace Health Saf.* 2016, 64, 28–36.

18. Atkinson R, and Arey J. Atmospheric chemistry of gas-phase polycyclic aromatic hydrocarbons: Formation of atmospheric mutagens. *Environ Health Perspect.* 1994, 102, 117–126.

19. Ramesh A, Archibong AE, Hood DB, Guo Z, and Loganathan BG. Global environmental distribution and human health effects of polycyclic aromatic hydrocarbons. In: Loganathan BG, and Lam PKS (eds) *Global Contamination Trends of Persistent Organic Chemicals.* CRC Press, Boca Raton, 2012, pp. 97–124.

20. Choi H, Harrison R, Komulainen H, Saborit JMD. Polycyclic aromatic hydrocarbons. In: *WHO Guidelines for Indoor Air Quality: Selected Pollutants.* World Health Organization, Copenhagen, 2010, pp. 289–345.

21. Davis E, Walker TR, Adams M, Willis R, Norris GA, and Henry RC. Source apportionment of polycyclic aromatic hydrocarbons (PAHs) in small craft harbor (SCH) surficial sediments in Nova Scotia, Canada. *Sci Total Environ.* 2019, 691, 528–537.

22. Valera B, Jørgensen ME, Jeppesen C, and Bjerregaard P. Exposure to persistent organic pollutants and risk of hypertension among Inuit from Greenland. *Environ Res.* 2013, 122, 65–73.

23. Langer P, Behzad N, Kočan A, Tajtáková M, Trnovec T, and Klimeš I. What we learned from the study of exposed population to PCBs and pesticides. *The Open Environ Pollut Toxicol J.* 2009, 1, 54–65.

24. Siddiqui MA, Laessig RH, and Reed KD. Polybrominated diphenyl ethers (PBDEs): new pollutants-old diseases. *Clin Med Res.* 2003, 1(4), 281–290.

25. Lamon L, Valle MD, Critto A, and Marcomini A. Introducing an integrated climate change perspective in POPs modelling, monitoring and regulation. *Environ Pollut.* 2009, 157, 1971–1980.

26. Johnson LL, Anulacion BF, Arkoosh MR, Burrows DG, da Silva DAM, Dietrich JP, Myers MS et al. Effects of legacy persistent organic pollutants (POPs) in fish – current and future challenges. In: Tierney KB, Farrell AP, and Brauner CJ (eds) *Organic Chemical Toxicology of Fishes*. Academic Press 2013, pp. 53–140.

27. Weber R, Watson A, Forter M, and Oliaei F. Persistent organic pollutants and landfills – a review of past experiences and future challenges. *Waste Manag Res*. 2011, 29(1), 107–121.

CHAPTER 10

Marine Litter Plastics and Microplastics and Their Toxic Chemicals Components
The Need for Urgent Preventive Measures*

Frederic Gallo, Cristina Fossi, Roland Weber, David Santillo, Joao Sousa, Imogen Ingram, Angel Nadal and Dolores Romano

CONTENTS

10.1 BACKGROUND—SITUATION ON PLASTIC AND RELATED CHEMICAL CONTAMINATION AND IMPACTS

10.1.1 Plastics in the Ocean: Sources, Volumes, Trends

Plastic marine litter is a mixture of macromolecules (polymers)* and chemicals, its size ranging from several meters to a few nanometers. It comprises such diverse items as fishing gear, agricultural plastics, bottles, bags, food packaging, taps, lids, straws, cigarette butts, industrial pellets, cosmetic microbeads, and the fragmentation debris coming from the weathering of all of them. It has become ubiquitous in all marine compartments, occurring on beaches, on the seabed, within sediments, in the water column and floating on the sea surface. The quantity observed floating in the open ocean represents only a fraction of the total input: over two-thirds of plastic litter ends up on the seabed, with half of the remainder washed up on beaches and the other half floating on or under the surface, so quantifying only floating plastic debris seriously underestimates the amounts of plastics in the oceans [1]. There are major concentration patches of floating plastics in all the five big ocean gyres, and there is evidence that even the polar areas are acting as additional global sinks of floating plastics [2].

The global production of plastics is following a clear exponential trend since the beginning of mass plastic consumption and production in the 1950s, and from a global production of 311 million tonnes in 2014, it is projected to reach around 1800 million tonnes in 2050 (Figure 10.1) [3]. The quantities of plastics leaking to the oceans on a global scale are largely unknown. Reliable quantitative estimations of input loads, sources and originating sectors represent a significant knowledge gap, but it is suggested that, every year, almost 8 million tonnes of plastic leak into the ocean. It is estimated that the ocean may already contain over 150 million tonnes of plastic [4], of which around 250,000 tonnes, fragmented into 5 trillion plastic pieces, may be floating at the ocean surface [5]. It has also been estimated that the global quantity of plastic in the ocean will nearly double to 250 million tonnes by 2025 [6],[†] which likely also represents a pollutant load of millions of tonnes of chemical additives. It is estimated that, on average, around 80%–90% of ocean plastic comes from land-based sources, including via rivers, with a smaller proportion arising from ocean-based sources such as fisheries, aquaculture and commercial cruise or private ships. Of that 80%, three-quarters is estimated to arise as a result of the lack of efficient collection schemes and proper waste management facilities in the municipalities in many countries, with the remainder entering the marine environment from careless littering and leaks from within the waste management system itself (such as urban drains).[‡]

In addition to the detrimental consequences that ingestion of plastics by marine biota may entail [8–10], worrying environmental consequences of marine litter also stem from microplastics (less than 5 mm in diameter) and nanoplastics (less than 100 nm in at least one of its dimensions), which could potentially affect marine biota both from their physical nature if ingested and by transfer of chemicals associated with them, including persistent organic pollutants (POPs) and endocrine disruptor chemicals (EDCs). Most micro- and nanoplastics originate from the degradation of macroplastics through different pathways, that is, photodegradation and other weathering processes of plastics that have leaked into

---

* Among the most common polymers found in the marine environment are low density polyethylene (PE-LD), linear low-density polyethylene (PE-LLD), high-density polyethylene (PE-HD), polypropylene (PP), polyethylene terephthalate (PET), polystyrene (PS) and polyvinyl chloride (PVC).

† The total estimated biomass of fish of 10 g per individual and upward in the oceans is 529 million tonnes [7], which puts the magnitude of the problem of plastics in the oceans into perspective.

‡ Other exogenous causes are natural disasters such as floods, hurricanes and tsunamis.

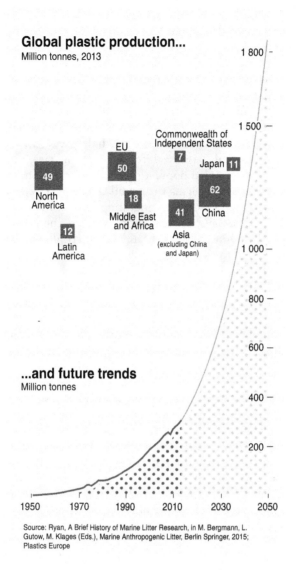

FIGURE 10.1 Global plastic production and future trends [3]; Marine Litter Vital Graphics- www.grida.no. Cartographer: Maphoto/Riccardo Pravettoni. (Source: UNEP. Marine plastic debris and microplastics—Global lessons and research to inspire action and guide policy change).

the sea [1] (e.g., bags, bottles, lids, food packaging, etc.), from plastic pellets lost into the environment during production or freight processes, or from textile fibers coming from washing machine runoff [3,11].* They may also be present as deliberately manufactured plastic microbeads used as scrubbing agents or for other purposes that can be found in some personal care and cosmetic products. It has been estimated that in the US alone, even considering that all sewage is connected to tertiary waste water treatment plants (WWTP), and assuming a 99% efficiency of the sedimentation process, around 8 trillion microbeads

---

* There are other sources of polymers that are not considered in this paper such as cigarette butts, tire and road wear and artificial turf infill.

may nevertheless be released into aquatic habitats every day. Furthermore, as the sludge of the WWTPs may subsequently be applied as fertilizer, part of the remaining 800 trillion microbeads may enter into soils and aquatic habitats via runoff [12].* Some wildlife may also contribute to the overall burden of microplastics when they ingest larger pieces of plastic, which are then broken up into smaller pieces in their guts and lost back into the environment in the form of microplastics. For example, fulmars (*Fulmarus glacialis*), a type of seabird, alone are estimated to reshape and redistribute annually about 6 tonnes of microplastics [13].

Uptake of microplastics through different mechanisms has been demonstrated in more than 100 marine species, from zooplankton to whales, including mussels, crabs, fish, planktivorous sharks, sea reptiles and seabirds. In some species, ingestion is reported in over 80% of individuals in sampled populations.† Organisms can ingest microplastics as food, whether unintentionally capturing them while filter- or deposit-feeding or mistaking them for prey when foraging, or even by ingesting prey containing microplastics, that is, trophic transfer [15]. In some species, microplastics can be taken into the body when they become entrapped by gill structures [16,17]. Microplastics and nanoplastics fall well within the size range of the staple phytoplankton diet of many zooplankton species, such as the Pacific krill. Fossi et al. [18] found that 56% of surface neustonic/planktonic samples from the Mediterranean Sea contained microplastic particles.

Microalgae attached to microplastics are assumed to be more easily captured by filter feeders than free microplastics in the water column [15]. After microplastics are assimilated into the organism, they accumulate in the gut, translocate into other tissues or are excreted, depending on the size, shape and composition of the particles. For example, fish fed with langoustines (*Nephrops norvegicus*) containing polypropylene filaments were found to ingest but not to excrete the microplastic strands, further corroborating the potential for trophic transfer and ecological impacts [14,19,20].

Uncertainties remain regarding the extent of harm caused to marine species directly by ingestion of microplastics and over the contribution they make to overall exposures to hazardous chemicals. Some studies report little or no physical or chemical harm to marine biota [21], while others including the use of the thermodynamic approach‡ and the simulation of physiological conditions in the gut, suggest that chemicals in plastics might be released to organisms after ingestion [22–25]. In mussels, *Mytilus galloprovincialis*, exposed to microplastics (polyethylene and polystyrene) contaminated with polyaromatic hydrocarbons, marked bioaccumulation of these chemicals was recorded in both the digestive gland and gills [26], similarly in tidal flat organisms such as lugworms, *Arenicola marina*, exposed to microplastics with adsorbed pollutants (nonylphenol and phenanthrene) and additive chemicals (Triclosan and PBDE-47) [24]. Endocytosis§ of plastic nanoparticles can also result in adverse toxic endpoints [1,19].

Microplastics move with currents, wave action and wind conditions and can be found throughout all marine compartments. Modeling the dynamics and fate of micro- and nanoplastics in the marine environment is a complex and uncertain task since particles initially at the sea surface can sink to sediments, accelerated by biofouling, ageing, etc., while those already in sediments can potentially become remobilized to the water column by bioturbation, resuspension or hydrodynamic conditions and translocation by marine organisms [15]. It is remarkable that benthic microplastics are far more widespread than previously assumed, with accumulation trends matching the increasing production of plastics worldwide [1,15,20].

---

* This was a strong argument for the law banning microbeads in cosmetics and personal care products in the US in 2015 (Microbead-Free Waters Act).
† Of the sampled crustacean *Nephrops norvegicus* in the Sea of Clyde (Scotland), 83% contained plastics (predominately filaments) in their stomachs [14].
‡ The study of transformations of matter and energy in systems as they approach equilibrium.
§ The taking in of matter by a living cell by invagination of its membrane.

In the Mediterranean Sea, marine litter has become a critical issue, as this is a region known to be accumulating a high concentration of plastics [27–29]. This is due to interaction of a number of factors, including the hydrodynamics of this semi-closed sea (from which outflow mainly occurs through deep water currents), combined with a lack or deficit of environmentally sound urban waste management and proper and efficient collection systems of much of the waste generated in many of its riparian countries and heavily populated coastal areas.

Other areas of particular concern include mid-ocean islands close to gyres and the Small Island Developing States (SIDS), where the situation has been depicted as "waste disaster" [30]. In addition to the challenge of marine litter, these States face serious deficiencies in basic waste management capabilities, due mainly to small and sparse populations with limited potential economies of scale. There is also a shortage of land for sanitary landfill, with waste often being disposed of casually by burial, burning or discard into the surrounding land and sea. Furthermore, consumption patterns are changing over time, with an increasing number of tourists and more plastic waste being generated overall. The state and pace of economic and social development in these small and remote countries, faced with growing populations and increasing urbanization and with limits to infrastructure and to both human and natural resources, make combatting this growing threat to their supporting ecosystems and means of life extremely challenging [3].

At a global level, UNEP has estimated the economic impact of marine plastics (excluding microplastics), including losses incurred by fisheries and tourism due to plastic littering, as well as beach clean-up costs, at around $13 billion per year [31].

10.1.2 Chemicals (POPs and EDCs) in Marine Litter Plastics: Fate in the Marine Environment

Besides the adverse physiological effects to marine organisms that arise from ingestion of pieces of plastic, plastics in the marine environment may also pose an additional chemical hazard, especially those containing known or suspected endocrine disrupting chemicals as additives or contaminants. Although plastics will not be the only route by which marine species are exposed to hazardous chemicals, existing evidence supports mounting concern in the scientific community that plastics may nonetheless make a significant contribution to exposures to complex mixtures of chemical contaminants [14,18,20,26,29,32–39,69]. The chemicals found in plastic marine litter can be classified in the following four categories of origin:

- Chemicals intentionally added during the production process (additives such as flame retardants, plasticizers, antioxidants, UV stabilizers and pigments);
- Unintentional chemicals coming from the production processes, including monomers (e.g., vinyl chloride, BPA, etc.)*—which may also originate from UV radiation onto the plastic waste—and catalysts, normally present in traces (ppm);
- Chemicals coming from the recycling of plastic waste†; and finally,
- Hydrophobic chemicals adsorbed from environmental pollution onto the surface of the plastics.‡

---

* Polymers can also be broken up into monomers by UV radiation, mechanical action, heat and other chemicals [40].
† Substances that were added intentionally in the virgin polymer and that are incorporated unknowingly or unwillingly when the plastic waste is recycled.
‡ Hydrophobicity is a property common to most of the POPs [41].

Whatever their origin, such substances may be directly released from plastics when they reach the guts of marine species, and may otherwise leach to the marine environment when the plastic weathers, at a rate depending on factors such as the nature and strength of the bond between additive and polymer (reactively bonded compounds requiring more energy), pore diameter, molecular weight of the additive, temperature, pressure and biofouling.

Chemicals with endocrine disrupting properties are a major concern for the marine environment. A compilation of lists of chemicals recognized as endocrine disrupting chemicals (EDCs) or suggested as potential EDCs has been developed by the International Panel on Chemical Pollution (IPCP) [42]. The SINList,* developed by ChemSec, compiles those chemicals with most urgent action needed.

In more general terms, experimental research on animals shows that low-level, non-linear exposures to endocrine disruptor chemicals (EDCs) lead to both transient and permanent changes to endocrine systems, as EDCs can mimic, compete with, or disrupt the synthesis of endogenous hormones [20,43,44]. This results in impaired reproduction and consequent low birth rates and potential loss of biodiversity, thyroid function and metabolism, and increased incidence and progression of hormone-sensitive cancers [45]. The research suggests that embryo and developmental periods are critical-sensitive periods to EDCs.[†] EDCs may cause effects in cellular and/or animal models at extremely low concentrations [45].

Some of those intentional chemical additives in plastics with toxic and endocrine-disrupting properties might be present at levels of 1000–500,000 mg/kg (ppm). This is the case of polybrominated diphenyl ethers (PBDEs) used as flame retardants in plastics, polyurethane foams and textiles; tetrabromobisphenol A (TBBPA)[‡] [40], used as flame retardant in epoxy, vinyl esters and polycarbonate resins; or hexabromocyclododecane (HBCDD), used in polystyrene foam (EPS/XPS) or di-2-ethylhexyl phthalate (DEHP) in PVC. It is also recognized that such chemicals can be found as particularly prominent contaminants in marine species collected from areas in which flame retardant-treated plastics are in use. For example, elevated HBCDD levels were found in oysters from aquaculture farms at which EPS/XPS buoys containing HBCDD were present [46]. The observation that high levels of the y-HBCDD isomer, which dominates commercial mixtures of this flame retardant [47,48], can be detected in fish in some European waters [49] indicates that direct exposure to technical HBCDD present in the polymer matrix can also be a relevant exposure pathway for fish, as well as the wider environmental exposure to the more stable α-HBCDD.

Further evidence that some POP chemicals are transferred to animal tissues directly from ingested plastic rather than from polluted prey, for example, arises from a study by Tanaka et al. [23] on short-tailed shearwaters that frequently ingest plastics that they mistake for food. These researchers focused on the presence of specific congeners of PBDEs present in the plastic but not commonly found in their prey (pelagic fish), confirming the presence of those congeners in both the fatty tissues of the birds and in the plastics found in their stomachs.

Other plastic additives of concern in the marine environment include chlorinated paraffins[§] [50] added as flame retardants; polychlorinated biphenyls (PCBs) and polychlorinated

---

* The SIN (Substitute It Now!) List, developed by ChemSec, identifies 32 EDCs of high concern that would require immediate action toward substitution and 14 more chemicals with ED properties and additional hazardous properties as well. http://chemsec.org/business-tool/sin-list/ (accessed March 2017).

† A fact that should be taken into account when assessing EDC effects in animal models.

‡ TBBPA degrades to bisphenol A and to TBBPA dimethyl ether. Bisphenol A and phthalates are rapidly metabolized once ingested but their concentration within the tissues varies between species for the same exposure.

§ Short-chained chlorinated paraffins are listed in Annex A (elimination) of the Stockholm Convention since May 2017.

naphthalenes (PCNs) included in PVC coatings/paints, and sometimes released as fine particles from abrasive blasting from (e.g., bridges into waters in tonnes scale*) [51,52] and per- and polyfluorinated compounds (PFCs)[†] [53,54]. Fluorinated polymers containing perfluorooctanesulfonic acid (PFOS) and perfluorooctanoic acid (PFOA) precursors used in some textile fibers and in paper and paperboard articles (i.e., fast-food packaging and paper plates, cups, etc.) to provide grease and water resistance [55], can become microplastics/fibers in the aquatic environment and release PFOS when degrading or ingested[‡] [56].

Other chemicals of concern include plastic additives with known or suspected endocrine disrupting properties, including alkylphenols (octylphenol and nonylphenol) used mainly as antioxidants, bisphenol A (BPA) present in polycarbonate plastics as trace monomer, phthalate esters—e.g., di(2-ethylhexyl) phthalate (DEHP), diisodecyl phthalate (DIDP), diisononyl phthalate (DINP) and butyl benzyl phthalate (BPP)—widely used as plasticizers in proportions up to 60% of the weight of a plastic to increase properties such as flexibility, transparency or longevity, and organotin compounds (based on methyl, butyl or octyl groups, such as tributyltin[§]) used as stabilizing additives in some PVC polymers. For example, Takada et al. [57] and Hirai et al. [58] analyzed a wide range of chemicals in marine plastics collected from urban and remote beaches and open oceans, including theoretically "non-persistent" additives such as alkylphenols (i.e., nonylphenol, octylphenol and BPA) that were detected in concentrations ranging from ng/g to µg/g in polyethylene and polypropylene debris.[¶] Moreover, a significant correlation has been demonstrated [18,60] among seven different phthalate esters (phthalates or PAEs) present in samples taken in the same area of microplastics, plankton and bubbler samples of different cetacean species.[**]

Some of these chemicals with endocrine disruptor properties may not qualify as "persistent" under the strict criteria of the Stockholm Convention, which requires in the screening criteria of its Annex D, evidence of its half-life in water, soil, sediments and air. Nevertheless, when present in a polymeric matrix in marine conditions, they may be potentially as harmful as officially recognized POPs in terms of behavior and consequences in the marine environment, as their presence is "topped-up" by the continuous flow of "fresh" plastic waste in river discharges, urban runoff and wastewater and associated sediments [41,61]. Their adsorption to microplastics, combined with the harsher environmental conditions of low temperature and salinity, combined also with low light and low oxygen content in subsurface waters and sediments, may also enhance their persistence in marine systems and their mobility and fluxes through all the compartments of the marine environment [15]. Sorption of contaminants in nanopores of plastics may further inhibit contaminant biodegradation [62]. Taking into account also that is very difficult or even impossible to establish a threshold of toxicity for many EDCs, as low dose

---

* PCBs and PCNs have been used to some extent as flame retardants in cables and other polymers including PVC coatings for corrosion protection. Such coatings are sometimes removed from bridges and dams by abrasive blasting and end up in rivers and the sea.

† http://greensciencepolicy.org/highly-fluorinated-chemicals/.

‡ PFCs in the environment can last for millions of years.

§ Marine paint containing tributyltin was forbidden by the International Convention on the Control of Harmful Anti-fouling Systems in Ships (entered into force in 2008), signed by most of the countries.

¶ Teuten et al. [22] tested the sorption uptake and desorption kinetics of HOCs in different polymers in laboratory conditions, showing that glassy polymers such as PVC exhibit larger sorption capacities and slower HOC release rates than rubbery polymers such as high-density polyethylene. Mato et al. [59] showed that polyethylene has higher affinity than polypropylene for HOCs.

** This finding suggests a new non-invasive method, which is to use the PAEs found in plankton as tracers of the exposure/ingestion in cetaceans or other endangered species.

effects and non-monotonic dose responses (NMDR) are common [44], the overall result would be that those substances in plastics in the marine environment may, through their widespread and pervasive distribution, present equivalent levels of concerns to those of recognized POPs. In this regard, such characteristics and evidence would allow equating EDCs in marine plastic waste with the defining properties of a POP under the Stockholm Convention. This is further discussed in "Contribution from the Stockholm Convention on Persistent Organic Pollutants," on potential measures for the consideration of the Stockholm Convention.

It should be noted that recycled plastic/polymers can also carry a high content of toxic chemicals carried over from their source plastics, and may also therefore contribute to chemical exposure of the marine environment* when they reach the ocean. The fact that much of the plastic waste collected for recycling is exported to countries with low legal requirements or technical capabilities on the control of the different types and concentrations of hazardous substances contained in the plastics† is an added source of concern, as the concentration of those toxic chemicals may increase in the recycled products.

With regard to the pollutants present in sea water and adsorbed onto the plastic surface, it has been estimated that fluxes of PCBs, PBDEs and PFOA to the Arctic caused by plastic debris was in the order of four to six times smaller than fluxes caused by atmospheric or seawater currents [63]. It is important to keep in mind, however, that the significance of pollutant transport routes does not only depend on the absolute amount of pollutants, but also on their impact from direct plastic ingestion and bioaccumulation in food chains [40]. In this regard, a qualitative distinction has to be made between microplastics and nanoplastics.

In microplastics, the adsorption of pollutants has been experimentally demonstrated from virgin plastic pellets in seawater, which implies that plastics constitute both a transport medium and a potential source of toxic chemicals in the marine environment [22,58,59]. The mechanisms of concentration of these chemicals is a complex issue depending of multiple variables including hydrophobicity of the pollutant, type of polymer, age of the plastic, water, temperature, pressure, presence of biofouling on the plastic surface and salinity. It is without doubt that other media present in the oceans, including natural sediments and the sorbent organic matter (SOM)—composed of suspended organic particulates, black carbon and natural diet and planktonic species—also have the capacity to adsorb hydrophobic organic chemicals (HOCs), such that ingestion of plastics will not be the only source of exposure to such chemical agents. Indeed, on average, the fraction of HOCs adsorbed to marine plastics appears to be statistically smaller when compared to that adsorbed fraction in other media in the ocean, such that chemical exposure of marine biota might be dominated by those other matrices [64]. Nonetheless, for certain chemical groups and/or specific local conditions with high concentrations of plastic matter, the importance of contaminant transfer from plastics may well be of quantitative significance.

In nanoplastics, the high surface area may present exceptionally strong sorption affinities for pollutants, thus changing the exposure and risk to these chemicals [65] and further increasing their significance as contributors to overall chemical exposure. In this regard, Koelmans et al. [66] affirm that: "because of the surface effect, it may be possible

---

* Articles with any substance listed under the Stockholm Convention, such as HBCDD used mainly in EPS/XPS polymers, are not allowed to undergo recycling processes, except articles (plastics) with hexa-, hepta-, tetra- or pentabromodiphenyl ethers that would allow some countries to recycle them until 2030 under an exemption of the Convention.
† For example, 50% of the plastic waste collected for recycling in the EU is exported to third countries with no sound environmental waste management guarantees (source: Plastic Recyclers Europe).

that nanoplastics retain organic toxic chemicals or heavy metals at higher concentrations than microplastics, thus leading to a fugacity gradient to organism tissue once ingested. If nanoplastics are capable of permeating membranes, passing cell walls, translocate and/or reside in epithelial tissues for prolonged times, the combination of particle and chemical toxicity may yield unforeseen risks." Velzeboer et al. [65] affirm that: "Nano- plastics have been shown to pass through the chorion of fish eggs and have been shown to move directly from the digestive tract of mussels into their circulatory system. This implies that occurrence of HOC contaminated nanoplastics in the environment may potentially enhance uptake."

Unfortunately, there are currently no sufficiently developed analytical methods adequate to detect and quantify nanoplastics in the environment or food chain [67], let alone to analyze their chemical signature in detail.

10.1.3 Potential Impacts on Marine Biodiversity

Chronic exposure simply to the physical presence of microplastics has been linked to effects on populations, including the negative influence of micro- and nanoplastics on survival and mortality of different species of zooplankton, which represent a critical energy source in the marine environment [68], or the reduced growth of offspring and reduced survival and fecundity compared with control organisms in crustaceans [10]. The Joint Research Centre of the EC [9] concluded that there is experimental evidence of negative physical/mechanical impacts from ingestion of plastic on the condition, reproductive capacity and survival of individual marine organisms. However, the evidence is restricted to laboratory experiments with organisms from lower trophic levels. These findings imply evidence of harm in natural populations, but quantifying the extent of this harm would be extremely challenging and the extent of harm caused by ingestion is likely to be underestimated, because necropsies have to be carried out. With regard to the chemical transfer of chemicals from plastics, there is still need of more studies for reliable estimates to be made as to the contribution to EDC exposure of marine species arising from microplastic or nanoplastics uptake, and this is a serious knowledge gap. There is already some scientific evidence suggestive of endocrine disruptor activity relating to the intake of chemicals associated with microplastics via the filter-feeding mechanisms of animals like mussels or baleen whales [18], or via the magnifying effect of the food chain in top predators such as the swordfish [69]. Although, in these studies mentioned, it could be questioned what the main source of phthalates is— water pollution, microplastics and/or food chain—the most plausible thesis is that water is not the main source of the pollution: phthalates in water are found in high concentrations only in coastal environments. In the case of the baleen whales, phthalate concentrations were very high in the microplastic and krill to which the animals were exposed, while not being detected in the water, though the relative contributions of krill and microplastics to overall phthalate exposure have yet to be determined.

While it is true that the transfer of persistent organic pollutants such as PCBs to aquatic organisms from microplastic in the diet is likely a small contribution compared to other natural pathways of exposure [70], this would not be the case for nonpersistent pollutants such as some EDCs, which are found in greater concentrations in microplastics than in surrounding seawater or sediments.

Widely used plasticizers with endocrine disrupting properties, for example, dibutyl phthalate, dimethyl phthalate, butyl benzyl phthalate or plastic monomers such as bisphenol A (BPA), can affect both development and reproduction in marine species: effect concentrations of plasticizers in laboratory experiments in some sensitive species such as mollusks, crustaceans and amphibians (including disturbance in spermatogenesis in fish) coincide with measured environmental concentrations in the low nanogram/liter to microgram/liter range. It should be remarked that there are still basic knowledge gaps, including the long-term exposures to environmentally relevant concentrations and their ecotoxicity when part of complex mixtures [61]. Other EDCs, such as alkylphenols, have the capacity to derail male reproductive development leading to feminization or demasculinization of the male form in fish and altered sex in mollusks. Others, such as tin-containing plastic stabilizers, elicit immunological disorders in fishes and induce imposex in gastropods [71].

10.1.4 Potential Impacts from Marine Plastics on Human Health

Although there are no current scientific studies correlating the direct consumption of fish or shellfish contaminated with microplastics containing or polluted with EDCs and the consequent endocrine disruption effects on human health, this is perhaps not surprising given the complexity of the issue [72,73]. One of the conclusions of the recent report of FAO on food safety [67] is that basic toxicological data on the consumption of micro- and nanoplastics in humans for a food risk safety assessment are essential lacking: the available data of toxicokinetics only include absorption and distribution, whereas no information is available on metabolism and little on excretion. It is not known whether or not ingested microplastics can be degraded into nanoplastics, and no data are available on the potential impact that cooking and/or processing seafood at high temperature may have on the toxicity of microplastics.

According to EFSA [74], a worst case estimate of exposure to microplastics after consumption of a portion of mussels (225 g) would be 7 µg of plastics. Based on this estimate and considering the highest concentrations of additives or contaminants reported in microplastics, and assuming complete release from microplastics, the microplastics will have a negligible effect on the total dietary exposure to persistent, bioaccumulative and toxic chemicals (PBT) and plastic additives, for example, in the case of bisphenol A (BPA), this would represent a contribution of less than 0.2% of the estimated dietary exposure to this compound in an adult of 70 kg.

With regard to existing evidence on the consequences of the uptake of micro- and nanoplastics by humans, medical literature on the impact of micro- and nanoplastics originating from inhalation or released from wear debris from plastic prosthetic implants shows diverse effects varying from DNA damage, changes in gene and protein expression, cell clotting, necrosis, apoptosis, proliferation and loss of cell viability, oxidative stress, increased Ca ions, inflammation and bone osteolysis, to lesions in organs [67].

However, at this time, the uncertainties surrounding potential health impacts remain high, and the data gaps, very large, including a lack of knowledge on the role and hazards of nanoplastics, potentially the most hazardous area of marine plastics [66,75]. Given the unavoidable increase in the coming decades of micro- and nanoplastics in the marine environment owing to the weathering and fragmentation of already existing "stocks" of marine macroplastics as well as future inputs, there is an urgency in better resolving the nature and scale of possible health effects, and in the meantime at least, to apply the precautionary principle.* Until the weight of the scientific evidence is more conclusive

---

* Precautionary principle by virtue of which where there are threats of serious or irreversible damage, lack of full scientific certainty shall not be used as a reason for postponing cost-effective measures to prevent environmental degradation.

regarding the risk that diets rich in small fish in whole (i.e., including the guts), or in bivalves and crustaceans containing microplastics or nanoplastics in significant quantities, could affect human endocrine systems—especially during embryo and infancy stages—or induce hepatic stress or other related health affections, it would seem wise to assume that measures that can limit or avoid intakes of microplastics would be an appropriate and important priority for public policy.

Further scientific research is needed with urgency on the potential impacts to endocrine systems and overall human health, especially on developing stages, by the direct or indirect ingestion of marine micro- and nanoplastics.

10.1.5 Potential Impacts on Food Safety and Availability and Economic Activity

Without immediate action, the environmental impacts and the economic costs are due to increase: as mentioned in the section "Plastics in the ocean: Sources, volumes, trends," more than a hundred million tonnes of plastics are estimated to have been dumped already to the oceans, and projections in plastic production and consumption indicate that plastic waste inputs in the sea may have an exponential increase if no urgent actions are taken [6]: on average, plastic consumption reached 100 kg per person per year in Western Europe and North America and 20 kg in Asia [76], and these figures are expected to grow rapidly in populated developing countries as urban population increases and urban dwellers must purchase all of their—plastic-packaged—food and beverage (see Figure 10.1).

As stated before, EDCs introduced via plastics may already be affecting marine biodiversity, raising additional concerns about food safety and security in the near future. Perhaps the most important source of dietary exposure of humans to microplastics at present is via filter-feeding shellfish, which retain particles from suspension on their gills for subsequent ingestion and thus they are directly exposed to micro- and nanoplastics via the water column. There is ample evidence of the ingestion of microplastics by bivalves [26], for example, nine of the most commercially popular species of bivalves purchased from a fishing market in Shanghai were found to be contaminated with microplastics. Based on the abundances observed, it was estimated that Chinese shellfish consumers could be exposed to 100,000s of microplastics each year [23,77].

In the case that marine biodiversity and food safety and availability are affected, this would represent a serious economic impact at a global level, especially in countries/islands where fish is a staple food, by exacerbating poverty [41,78,79] in a context of climate change and growing competition for natural resources. Fish contributes, or exceeds, 50% of total animal protein intake in some Small Island Developing States, as well as in Bangladesh, Cambodia, Ghana, Indonesia, Sierra Leone and Sri Lanka [80]. It is estimated that fish, bivalves and crustaceans provide more than 3.2 billion people with almost 20% of their average per capita intake of animal protein, and 5.1 billion people with 10% of such protein. Over 53% of the global trade in fish and seafood originates in developing countries whose net trade income (export–import), valued at US$35 billion in 2012, is greater than the net trade income of the other agricultural commodities combined. Furthermore, around 260 million people are involved in global marine capture fisheries, including full-time and part-time jobs in the direct and indirect sectors [67,81].

As a reference for the economic magnitude of the problems posed by "on land" endocrine disruptor chemicals, according to a series of studies released by the Endocrine

Society, and only taking into account medical costs,* routine exposure to EDCs found in pesticides and in everyday consumer items in homes costs, only to the EU, €157 billion annually [82] and $340 billion annually in the US [83], a magnitude similar to the cost of smoking-related illness—the largest single cost coming from effects on children.

10.2 CONCLUSIONS—ACTIONS NEEDED AND POTENTIAL SUPPORT BY CHEMICAL AND WASTE CONVENTION

10.2.1 Urgent Measures Needed on Production and Consumption of Plastics and Waste Management

One urgent measure would be global fully fledged efficient waste collection, management, recycling and environmentally sound disposal systems that would guarantee an almost zero plastic release to the environment. However, this seems a financially challenging and possibly decades-long endeavor. Moreover, while such an infrastructure could be economically feasible in industrial countries, it may not be feasible or cost-effective for developing nations [84]. In addition, the exponentially increasing global trend of plastic production and consumption, in a context of global financial crisis, makes extremely uncertain the ability to achieve already established objectives of reduction of marine litter† at global, regional, subregional or national levels. Furthermore, the more frequent and strong flooding events in the different world regions facilitate the flushing of plastic to waters in developing countries but also in industrial countries since plastic waste just get flushed away.

Therefore, urgent and strong actions with relatively low public investment are needed at global level, that is, policy reforms including extended producer responsibility (EPR) and fiscal and economic instruments. A prevention and "best available techniques and practices" approach, built on a holistic life cycle basis, could allow scarce resources and effort to be focused on measures that are very likely to reduce the problem by directly attacking the source, similar to the way in which industrial toxic emissions were effectively curbed in some developed countries at the end of the last century, instead of relying on "end-of-pipe" solutions, for example, focusing only on cleaning measures such as "fishing for—floating macro—plastic," which are not efficient and economically viable in an oceanic scale‡ and which do not stop the continuous inputs of plastic, the already existing microplastic pollution or sunk plastics or by only assessing and monitoring how much worse the problem is getting [86].

Although there is still need to carry out focused scientific research to fill the knowledge gaps about the impacts of plastic litter in the marine environment [87], the food chain and human health, the precautionary principle, the already existing scientific evidence and reasonable concerns should be enough to support actions by the scientific, industry, policy

---

* The Endocrine Society has recently stated that: "... data reviewed in EDC-2 removes any doubt that EDCs are contributing to increased chronic disease burdens related to obesity, diabetes mellitus, reproduction, thyroid, cancers, and neuroendocrine and neurodevelopmental functions" [45].

† The Honolulu Strategy, the global framework to prevent marine litter, does not prescribe specific marine debris reduction targets but expects "substantial progress" by 2030. The UN's Sustainable Development Goal number 14 (Sustainable Oceans) aims to "prevent and significantly reduce" marine litter in 2025. In the European Union, a 30% reduction for beach litter by 2025, compared with 2015 levels, has been proposed for all its regional seas.

‡ Clean-up may be a suitable last resort for addressing marine litter in limited zones such as urban areas, tourist beaches and ports where the litter causes severe social and economical damage [85], or in marine special protected areas (SPAs).

and civil society communities to curb the leaking of plastics into the marine environment in the short term. To think in terms of "business as usual" and "adaptation measures" to cope with plastic pollution in the oceans instead of prevention and mitigation measures would lead to another predictable environmental crisis for future generations to cope with. The dangers of working in isolation are already apparent from industry-centered responses such as the development of "oxodegradable" plastic products, which merely take out of sight plastics by fragmenting them at the end of their lifetime into numerous small but essentially nondegradable pieces [84].

Strong policy actions to curb unnecessary plastic packaging on the demand side on the short term, such as the ban on free single-use plastic bags, or to substantially increase the collection rate of plastic waste, such as the deposit-refund schemes for plastic beverage bottles,* which have a demonstrated high rate of success in many countries,† and the ban on plastic microbeads in cosmetics and personal care products, are strongly needed at regional, subregional or national levels as part of their strategies for waste management. Initiatives to promote measurement of the types and quantities of plastic used by companies or communities, such as the "Plastic Disclosure Project,"‡ could facilitate accountability and the implementation of measures to reduce avoidable plastic use by the private and public sectors. Designers and producers should avoid creating products that are inherently single use or inevitably destined for landfill [85].

Other measures to consider in developing countries or remote rural communities of Africa, America or Pacific SIDS, with no or few environmentally sound disposal facilities, would be, for example, the take back or repatriation schemes of plastic waste under extended producer responsibility (EPR) schemes, especially for food and beverage plastic packaging, given the clear benefits of plastic versus other packaging materials in reducing the total amount of packaging (in tonnes), as well as the energy required for transportation on the long-haul shipments and the food losses.

Campaigns to make plastic litter socially unacceptable and educate consumers across the supply chain would be necessary elements of any policy of awareness on waste. Designing for recycling would allow to divert important volumes of plastic waste from the waste management systems. It is necessary to work with companies and research institutes, especially in the food sector, to optimize food packaging and materials to avoid unnecessary use of persistent plastics and toxic chemicals. Strong policy actions, as well as more research, development and innovation in green chemistry are needed for the substitution of POPs, EDC and other toxic substances in plastics as well as for the development of more benign alternatives to persistent polymers in the marine environment.

It is important to highlight that compostable bioplastics or plastics labeled as "biodegradable in the environment" are not degraded in marine conditions, where parameters such as temperature, oxygen, and salinity are very different than those expected in a composting process, and so they have equivalent properties in the marine environment in this regard as persistent plastics.§ Other innovative materials, such as marine biodegradable polymers, especially for food packaging, could have an important role to play in reducing the

---

* Plastic beverage bottles represent around 20% of all plastic packaging waste in the EU.
† Compared with the relatively low and stagnate rates of curbside separate collection of plastic packaging waste, with the added benefit of delivering a high-quality product ready for recycling [88].
‡ http://Plasticdisclosure.org.
§ Biodegradation according to EN13432 is considered to be complete if at least 90% of the material has been converted into carbon dioxide (the remainder is due to the fact that besides carbon dioxide, water and biomass are produced during biodegradation).When all the organic carbons in the polymer are converted, it is referred as complete mineralization.

environmental damage of plastics leaking to the marine environment, but the biodegradability in the marine environment of such alternative plastics (such as the polyhydroxyalkanoates, PHAs) would require further study and validation under a range of conditions in seawater, and internationally accepted certification seals. Further avenues of research on these biomaterials would be to study their complete lifecycle (e.g., to ensure that they do not compete with food production, best options to recycle), potential harms by ingestion to marine biota, and its rate of adsorption of HOC in seawater before its degradation compared with other adsorbing media in the marine environment, including persistent plastics.

Implementing or improving environmentally sound waste collection and management systems of urban waste represents a basic necessary step to reducing plastic inputs, especially in developing economies. Special attention should be paid to avoid creating further environmental and health impacts, for example, by promoting non-best available technology (BAT) waste incineration of plastics without tight environmental controls, which may be an important identified source of POPs, such as dioxins and furans. Effective mandatory or voluntary measures are urgently needed to curb the consumption of single-use plastics, as well as the urgent banning of microplastics in all types of cosmetics and personal care products, even in those countries with 100% coverage of tertiary WWTP.

The actual levels of POPs in marine plastics collected from the sea should be taken into consideration when deciding on management options for marine waste, including recycling.

The implementation of action plans to reduce the input of marine plastic around the world needs to involve all stakeholders from the local and national authorities to international bodies, the scientific community, plastic manufacturers and retailers, tourism and fishing industries, NGOs, etc., to effectively address socioeconomic and environmental issues related to plastic pollution from a sustainable and global point of view [89].

10.2.2 Potential Measures Suggested in the Framework of the Stockholm and Basel Conventions to Address Marine Litter; Contribution from the Stockholm Convention on Persistent Organic Pollutants (POPs)

Plastic marine litter is an issue of global environmental and health concern, due to its persistence, wide geographical distribution and long-range transport capacity of persistent and toxic chemicals in the marine environment.

Due to the toxic chemical exposure of marine biota through marine plastic litter and the related bioaccumulation and widespread distribution in all marine compartments of persistent micro- and nanoplastics with chemicals of concern acting as persistent organic pollutants in the marine environment and given the potential human affection to consider:

1. To take into account the risks of additives in plastics with endocrine disruptor properties when selecting and assessing substances for the listing of new POPs in the Stockholm Convention. Some plastic additives with endocrine disruptive properties which might not pass some of the POPs screening criteria such as persistence in water in standard laboratory conditions, are expected to have longer half-life in the plastic due to the protection (or molecular encapsulation) within the polymer matrix, and may have even longer half-life in the marine environment, due to its physical and chemical properties such as lower temperatures, lower oxygen levels, salinity, pH and lower levels of light in water column and sea floor and sediments, i.e. theoretically "non-persistent" chemical additives or trace monomers in plastics (such as alkylphenols, phthalates, BPA) have been detected in high concentrations in floating polyethylene and polypropylene plastic—the

most widely used in packaging—in open oceans [18,58,60,69]. In addition, apart from their mobility and fluxes through all the compartments of the marine environment [15], the new inputs of "fresh" plastic into the marine environment are so continuous and widespread through all the oceans that would be equivalent to the continental or oceanic long-range transport property of highly persistent POPs. Their exposure to marine biota is relevant because

a. The very low doses of EDCs required to affect the endocrine systems in marine biota and humans [90], compared to those required in toxicological tests to prove carcinogenicity in candidate POPs, especially during the embryo and developing stages,

b. The uptake of microplastics containing those chemicals by marine biota, which may affect biodiversity, food security, food availability and potentially human health, especially if the persistent plastic consumption and production follows the expected growing trends in the coming decades (see Figure 10.1), without the necessary environmentally sound waste management and collection facilities being in place globally to avoid plastic leaking into the oceans.

2. The introduction of measures to reduce marine plastic litter in National Implementation Plans for the Stockholm Convention on Persistent Organic Pollutants, such as

a. Promoting substitution and green chemistry to avoid POPs and other harmful chemicals in plastics, especially EDCs.

b. Encouraging plastic waste prevention and supporting development and implementation of safer or more benign alternatives to persistent plastics in the marine environment.

c. Supporting research on environmental and health impacts of marine plastics, microplastics and nanoplastics and related fate of EDCs and POPs.

d. Encouraging ecodesign for better packaging recyclability.

e. Encouraging plastic waste recycling when feasible.

f. Promoting BATs to reduce plastic leakage to oceans and improving information on input loads, sources and originating sectors.

g. Encouraging the improvement and efficiency of collection and sound environmental management of waste.

h. Encouraging changes in consumption and littering behavior.

10.2.3 Contribution from the Basel Convention on Hazardous Wastes

To acknowledge plastic marine litter as an issue of global environmental and health concern, due to its persistence, wide geographical distribution and long-range transport capacity of toxic chemicals in the marine environment and the need to address it by improvement of waste management and other means.

To consider:

1. To include measures to avoid or reduce marine plastic litter in the Strategic Framework for the implementation of the Basel Convention.

2. Revising Annexes I and III of the Convention to ensure the listing of all chemicals with endocrine disruptor substances (EDCs) in plastics that may end up as microplastic waste in the marine environment.

3. The adoption of new guidelines on environmental sound management of plastic and plastic containing wastes, with a view to minimize the possibility of plastic leaks into the oceans coming from waste management.

4. Reviewing policies related to the export of plastic containing waste to countries where no environmentally sound recycling, recovery or final disposal of the plastic materials contained in the waste are guaranteed, i.e. uncontrolled recycling of plastics with toxic chemicals, waste disposal in non-BAT open dumps, or incinerated in cement furnaces with no environmental controls, or non-BAT incinerators without tight environmental measures and controls like dioxin catalyzers, continuous outflow monitoring and sound environmental landfilling of its ashes.

5. Ensuring the best available techniques and best environmental practices are recommended in Basel Convention waste guidelines and manuals to avoid disposal methods that might re-release toxic chemicals into the air, water or soils to safeguard the health of neighboring communities.

6. Developing efficient strategies for achieving the prevention and minimization of the generation of marine plastic litter.

10.3 FUTURE ACTIVITIES TO ADDRESS MARINE LITTER

The Working Group identified a number of possible future activities to address the issue by the Basel and Stockholm Conventions Regional Centres in coordination with existing platforms, or by any other UN Environment institutions, IGOs, governments, NGOs, etc., such as

- Dissemination, information and training activities to improve awareness and knowledge on the risks and challenges posed by marine plastic litter and on measures to combat it.
- Technical assistance and capacity-building activities to support parties and other stakeholders in implementing waste management and efficient waste collection measures to reduce plastic marine litter.
- Develop recommendations to review regional and national regulatory frameworks concerning plastic and plastic containing wastes and inclusion of measures to prevent plastic waste, such as measures to reduce plastic bags consumption and establishment of "deposit and return" schemes for beverage packaging.
- To promote innovation and technology transfer to avoid persistent plastics and sound chemical substitution of toxic components in plastic packaging and other plastics, encouraging plastic waste prevention and supporting development and implementation of safer or more benign alternatives to persistent plastics in the marine environment.
- To assist developing countries, economies in transition and Small Island Developing States with efficient collection and environmentally sound management of plastic waste and plastic packaging, which they are unable to dispose of or recycle in an environmentally sound manner but continue to receive nonetheless, including through take back or repatriation policies under extended producer responsibility (EPR) schemes.

REFERENCES

1. Andrady A. Microplastics in the marine environment. *Mar Pollut Bull* 2011, 62(8), 1596–1605.
2. Van Sebille E, England MH, and Froyland G. Origin, dynamics and evolution of ocean garbage patches from observed surface drifters. *Environ Res Lett* 2012, 7, 044040.

3. UNEP. *Marine Plastic Debris and Microplastics. Global Lessons and Research to Inspire Action and Guide Policy Change.* United Nations Environment Programme, Nairobi, 2016.

4. McKinsey Center for Business and Environment. Stemming the tide: land-based strategies for a plastic-free ocean. McKinsey & Company and Ocean Conservancy, 2015.

5. Eriksen M, Lebreton LCM, Carson HS, Thiel M, Moore CJ, Borerro JC, Galgani F, Ryan PG, and Reisser J. Plastic pollution in the world's oceans: More than 5 trillion plastic pieces weighing over 250 000 tons afloat at Sea. *PLoS ONE* 2014, 9(12), e111913.

6. Jambeck JR, Geyer R, Wilcox C, Siegler TR, Perryman M, Andrady A, Narayan R, and Law KL. Plastic waste inputs from land into the ocean. *Science* 2015, 347(6223), 768–771.

7. Jennings S, Mélin F, Blanchard JL, Forster R, Dulvy NK, Wilson RW. Global-scale pre- dictions of community and ecosystem properties from simple ecological theory. *Proc Royal Soc* 2008, 275, 1375–1383.

8. Wrighta SL, Thompson RC, and Galloway TS. The physical impacts of microplastics on marine organisms: A review. *Environ Pollut* 2013, 178, 483–492.

9. Werner S, Budziak A, van Franeker J, Galgani F, Hanke G, Maes T, Matiddi M et al. *Harm caused by marine litter.* MSFD GES TG marine litter— Thematic Report; JRC Technical report; EUR 28317 EN. 2016.

10. Anderson JC, Park BJ, and Palace VP. Microplastics in aquatic environments: Implications for Canadian ecosystems. *Environ Pollut* 2016, 218, 269–280.

11. Duis K and Coors A. Microplastics in the aquatic and terrestrial environment: Sources (with a specific focus on personal care products), fate and effects. *Environ Sci Eur* 2016, 2016(28), 2.

12. Rochman CM, Kross SM, Armstrong JB, Bogan MT, Darling ES, Gren SJ, Smyth AR et al. Scientific evidence supports a ban on microbeads. *Environ Sci Technol* 2015, 49(18), 10759–10761.

13. Van Franeker JA and Meijboom A. *Litter NSV—Marine Litter Monitoring by Northern Fulmars: A Pilot Study.* Alterra, Provo, 2002.

14. Murray F and Cowie PR. Plastic contamination in the decapod crustacean *Nephrops norvegicus. Mar Pollut Bull* 2011, 62, 1207–1217.

15. GESAMP. Sources, fate and effects of microplastics in the marine environment: Part two of a global assessment. In: Kershaw PJ, and Rochman CM (eds) *IMO/FAO/ UNESCO-IOC/UNIDO/WMO/IAEA/UN/UNEP/UNDP Joint Group of Experts on the Scientific Aspects of Marine Environmental Protection Rep. Stud. GESAMP No. 93.* 2016, p. 220.

16. Watts A, Lewis C, Goodhead RM, Beckett SJ, Moger J, Tyler CR, and Galloway TS. Uptake and retention of microplastics by the shore crab *Carcinus maenas. Environ Sci Technol* 2014, 48(15), 8823–8830.

17. Fernández P, Leslie H, and Ferreira M (eds). The CleanSea project: An interdisciplinary study of marine litter in the EU. Special issue 'Coastal & Marine' magazine, vol 2015-1. 2015.

18. Fossi MC, Panti C, Guerranti C, Coppola D, Giannetti M, Marsili L, and Minutoli R. Are baleen whales exposed to the threat of microplastics? A case study of the Mediterranean fin whale. *Mar Pollut Bull* 2012, 64(11), 2374–2379.

19. GESAMP. Sources, fate and effects of microplastics in the marine environment: A global assessment. In: Kershaw PJ (ed) *IMO/FAO/UNESCO- IOC/UNIDO/*

WMO/IAEA/UN/UNEP/UNDP Joint Group of Experts on the Scientific Aspects of Marine Environmental Protection Rep. Stud. GESAMP No. 90. 2015, p. 96.

20. Avio CG, Gorbi S, and Regoli F. Plastics and microplastics in the oceans: From emerging pollutants to emerged threat. *Mar Environ Res* 2017, 128, 2–11.

21. Koelmans AA, Besseling E, and Foekema EM. Leaching of plastic additives to marine organisms. *Environ Pollut* 2014, 187, 49–54.

22. Teuten EL, Saquing JM, Knappe DR, Barlaz MA, Jonsson S, Björn A, Rowland SJ et al. Transport and release of chemicals from plastics to the environment and the wildlife. *Phil Trans R Soc B* 2009, 364, 2027–2045.

23. Tanaka K, Takada H, Yamashita R, Mizukawa K, Fukuwaka MA, and Watanuki Y. Accumulation of plastic-derived chemicals in tissues of seabirds ingesting marine plastics. *Mar Pollut Bull* 2013, 69(1–2), 219–222.

24. Browne MA, Niven SJ, Galloway TS, Rowland SJ, and Thompson RC. Microplastic moves pollutants and additives to worms, reducing functions linked to health and biodiversity. *Curr Biol* 2013, 23, 2388–2392.

25. Bakir A, Rowland SJ, and Thompson RC. Enhanced desorption of persistent organic pollutants from microplastics under simulated physiological conditions. *Environ Pollut* 2014, 185, 16–23.

26. Avio CG, Gorbi S, Milan M, Benedetti M, Fattorini D, d'Errico G, Pauletto M, Bargelloni L, and Regoli F. Pollutants bioavailability and toxicological risk from microplastics to marine mussels. *Environ Pollut* 2015, 198, 211–222.

27. Cózar A, Sanz-Martín M, Martí E, González-Gordillo JI, Ubeda B, Gálvez JÁ et al. Plastic accumulation in the Mediterranean Sea. *PLoS ONE* 2015, 10, e0121762.

28. UNEP/MAP. *Marine Litter Assessment in the Mediterranean 2105.* United Nations Environment Programme/Mediterranean Action Plan (UNEP/MAP), Nairobi, 2015.

29. Fossi MC, Romeo T, Baini M, Panti C, Marsili L, Campani T, Canese S et al. Plastic debris occurrence, convergence areas and Fin Whales feeding ground in the Mediterranean Marine Protected Area Pelagos Sanctuary: A modelling approach. *Front Mar Sci* 2017, 4, 167.

30. Veitayaki J. Pacific Islands drowning in their waste: Waste management issues that threaten sustainability. Proceedings of international seminar on islands and oceans. Ocean Policy Research Foundation, Nippon Foundation, 2010.

31. UNEP. *Valuing Plastics: The Business Case for Measuring, Managing and Disclosing Plastic use in the Consumer Goods Industry.* United Nations Environment Program, Nairobi, 2014.

32. Rochman CM, Hoh E, Kurobe T, and The SJ. Ingested plastic transfers hazardous chemicals to fish and induces hepatic stress. *Sci Rep* 2013, 3, 3263.

33. Rochman CM. Plastics and priority pollutants: A multiple stressor in aquatic habitats. *Environ Sci Technol* 2013, 2013(47), 2439–2440.

34. Colton JB, Knapp FD, and Burns BR. Plastic particles in surface waters of the Northwestern Atlantic. *Science* 1974, 185, 491–497.

35. Ng KL and Obbard JP. Prevalence of microplastics in Singapore's coastal marine environment. *Mar Pollut Bull* 2006, 52, 761–767.

36. Rios LM, Moore C, and Jones PR. Persistent organic pollutants carried by synthetic polymers in the ocean environment. *Mar Pollut Bull* 2007, 54, 1230–1237.

37. Fossi MC, Marsili L, Baini M, Giannetti M, Coppola D, Guerranti C, Caliani L et al. Fin whales and microplastics: The Mediterranean Sea and the Sea of Cortez scenarios. *Environ Pollut* 2016, 209, 68–78.

38. UNEP. *Marine Litter, an Analytical Overview.* United Nations Environment Program, Nairobi, 2005.

39. Fossi MC, Baini M, Panti C, Galli M, Jiménez B, Muñoz-Arnanz J, Marsili L, Finoia MG, and Ramírez-Macías D. Are whale sharks exposed to persistent organic pollutants and plastic pollution in the Gulf of California (Mexico)? First ecotoxicological investigation using skin biopsies. *Comp Biochem Physiol* 2017, 199, 48–58.

40. Science for Environment Policy. *Plastic Waste: Ecological and Human Health Impacts. DG Environment News Alert Service.* European Commission, Brussels, 2011.

41. Nerland IL, Halsband C, Allan I, and Thomas KV. *Microplastics in Marine Environments: Occurrence, Distribution and Effects.* Norwegian Institute for Water Research, Kristiansand, 2014.

42. International Panel on Chemical Pollution (IPCP). Overview Report I: A compilation of lists of chemicals recognised as endocrine disrupting chemicals (EDCs) or suggested as potential EDCs. 2016. http://wedocs.unep. org/handle/20.500.11822/12218. Accessed February 2017.

43. Talsness CE, Andrade AJM, Kuriyama SN, Taylor JA, and vom Saal FS. Components of plastic: Experimental studies in animals and relevance for human health. *Phil Trans R Soc B* 2009, 364, 2079–2096.

44. Munn S and Goumenou M. Thresholds for endocrine disrupters and related uncertainties. Report of the Endocrine Disrupters Expert Advisory Group (ED EAG). European Commission, Joint Research Centre. Institute for Health and Consumer Protection, 2013.

45. Gore AC, Chappell VA, Fenton SE, Flaws JA, Nadal A, Prins GS, Toppari J, and Zoeller RT. EDC-2: The endocrine society's second scientific state- ment on endocrine-disrupting chemicals. *Endocr Rev* 2015, 36(6), E1–E150.

46. Hong SH, Jang M, Rani M, Han GM, Song YK, and Shim WJ. Expanded polystyrene (EPS) buoy as a possible source of hexabromocyclodode- canes (HBCDs) in the marine environment. *Organohalogen Compd* 2013, 75, 882–885.

47. Nakagawa R, Murata S, Ashizuka Y, Shintani Y, Hori T, and Tsutsumi T. Hexabromocyclododecane determination in seafood samples collected from Japanese coastal areas. *Chemosphere* 2010, 81, 445–452.

48. Becher G. The stereochemistry of 1,2,5,6,9,10-hexabromocyclodo-decane and its graphic representation. *Chemosphere* 2005, 58, 989–991.

49. Rüdel H, Müller J, Quack M, and Klein R. Monitoring of hexabromo- cyclododecane diastereomers in fish from European freshwaters and estuaries. *Environ Sci Pollut Res* 2012, 19, 772–783.

50. Zhang Q, Wang J, Zhu J, Liu J, Zhang J, and Zhao M. Assessment of the endocrine-disrupting effects of short-chain chlorinated paraffins in *in vitro* models. *Environ Int* 2016, 94, 43–50.

51. Jartun M, Ottesen RT, Steinnes E, and Volden T. Painted surfaces—important sources of polychlorinated biphenyls (PCBs) contamination to the urban and marine environment. *Environ Pollut* 2009, 157(1), 295–302.

52. ELSA. *PCB in der Elbe –Eigenschaften, Vorkommen und Trends sowie Ursachen und Folgen der erhöhten Freisetzung im Jahr 2015.* Behörde für Umwelt und Energie Hamburg, Projekt Schadstoffsanierung Elbsedimente, Hamburg, 2016.

53. Wang Z, DeWitt J, Higgins C, and Cousins I. A Never-ending story of per- and polyfluoroalkyl substances (PFASs)? *Environ Sci Technol* 2017, 51(5), 2508–2518.

54. Washington JW, Ellington J, Jenkins TM, Evans JJ, Yoo H, and Hafner SC. Degradability of an acrylate-linked, fluorotelomer polymer in soil. *Environ Sci Technol* 2009, 43(17), 6617–6623.

55. Schaider L, Balan S, Blum A, Andrews D, Strynar M, Dickinson M, Lunderberg D, Lang J, and Peaslee G. Fluorinated Compounds in U.S. Fast Food Packaging. *Environ Sci Technol Lett* 2017, 4, 105–111.

56. Guerranti C, Ancora S, and Bianchi N. Perfluorinated compounds in blood of *Caretta caretta* from the Mediterranean Sea. *Mar Pollut Bull* 2013, 73, 98–101.

57. Takada H, Hirai H, Ogata Y, Yuyama M, Mizukawa K, Yamashita R, Watanuki Y et al. Global distribution of organic micropollut- ants in marine plastics. Non-published paper. 2011.

58. Hirai H, Takada H, Ogata Y, Yamashita R, Mizukawa K, Saha M, Kwan C et al. Organic micropollutants in marine plastic debris from the open ocean and remote and urban beaches. *Mar Pollut Bull* 2011, 62(8), 1683–1692.

59. Mato Y, Takada H, Zakaria MP, Kuriyama Y, and Kanehiro H. Toxic chemi- cals contained in plastic resin pellets in the marine environment—spatial difference in pollutant concentrations and the effects of resin type. *Kankyo Kagakukaishi* 2002, 15, 415–423.

60. Baini M, Martellini T, Cincinelli A, Campani T, Minutoli R, Panti C, Finoiac MG, and Fossia MC. First detection of seven phthalate esters (PAEs) as plastic tracers in superficial neutonic/planktonic samples and cetacean bubblers. *Analitical Methods* 2017, 9, 1512–1520. https://doi.org/10.1039/c6ay02674e

61. Oehlmann J, Schulte-Oehlmann U, Kloas W, Jagnytsch O, Lutz I, Kusk KO, Wollenberger L et al. A critical analysis of the biological impacts of plasticizers on wildlife. *Phil Trans R Soc B* 2009, 364, 2047–2062. https://doi.org/10.1098/rstb.2008.0242

62. Hatzinger PB and Alexander M. Biodegradation of organic compounds sequestered in organic solids or in nanopores within silica crystals. *Environ Toxicol Chem* 1997, 16, 2215–2221.

63. Zarfl C and Matthies M. Are marine plastic particles transport vectors for organic pollutants to the Arctic? *Mar Pollut Bull* 2010, 60, 1810–1814.

64. Koelmans A, Bakir A, Allen G, and Janssen C. Microplastic as a vector for chemicals in the aquatic environment: Critical review and model- supported reinterpretation of empirical studies. *Environ Sci Technol* 2016, 50(7), 3315–3326.

65. Velzeboer I, Kwadijk C, and Koelmans AA. Strong sorption of PCBs to nanoplastics, microplastics, carbon nanotubes, and fullerenes. *Environ Sci Technol* 2014, 48, 4869–4876.

66. Koelmans AA, Besseling E, and Shim WJ. Nanoplastics in the aquatic environment. Critical review. In: Bergmann M, Gutow L, and Klages M (eds) *Marine Anthropogenic Litter*. Springer, Berlin, 2015, p. 2015.

67. Lusher A, Hollman P, and Mendoza-Hill J. Microplastics in fisheries and aquaculture. Status of knowledge on their occurrence and implications for aquatic organisms and food safety. FAO Fisheries and Aquaculture Technical Paper, 2017, p. 615.

68. Desforges JPW, Galbraith M, and Ross PS. Ingestion of microplas- tics by zooplankton in the Northeast Pacific Ocean. *Arch Environ Contam Toxicol* 2015, 69, 320–330.

69. Fossi MC, Casini S, Ancora S, Moscatelli A, Ausili A, and Notarbartolo-di-Sciara G. Do endocrine disrupting chemicals threaten Mediterranean swordfish? Preliminary results of vitellogenin and Zona radiata proteins in *Xiphias gladius*. *Mar Environ Res* 2001, 52(5), 477–483.

70. Beckingham B and Ghosh U. Differential bioavailability of polychlorinated biphenyls associated with environmental particles: Microplastic in comparison to wood, coal and biochar. *Environ Pollut* 2017, 220, 150–158.

71. Bergman A, Heindel JJ, Jobling S, Kidd KA, and Zoeller T (eds). *State of the science of endocrine disrupting chemicals—2012*. WHO/UNEP, Geneva, 2013.

72. Wright SL and Kelly FJ. Plastic and human health: A micro issue? *Environ Sci Technol* 2017, 2017(51), 6634–6647.

73. Miller K, Santillo D, and Johnston P. *Plastics in sea food.* Greenpeace Research Laboratories Technical Report (Review) 05-2016. 2016.

74. EFSA CONTAM Panel (EFSA Panel on Contaminants in the Food Chain). Statement on the presence of microplastics and nanoplastics in food, with particular focus on seafood. *EFSA J* 2016, 14(6), 4501.

75. Bouwmeester H, Hollman PCH, and Peters RJB. Potential health impact of environmentally released micro- and nanoplastics in the human food production chain: Experiences from nanotoxicology. *Environ Sci Technol* 2015, 49, 8932–8947.

76. Gourmelon G. *Global Plastic Production Rises, Recycling Lags.* Worldwatch Institute, Washington, 2015.

77. Li J, Yang D, Li L, Jabeen K, and Shi H. Microplastics in commercial bivalves from China. *Environ Pollut* 2015, 207, 190–195.

78. McKinley A and Johnston EL. Impacts of contaminant sources on marine fish abundance and species richness: A review and meta-analysis of evidence from the field. *Mar Ecol Prog Ser* 2010, 420, 175–191.

79. Johnston EL and Roberts DA. Contaminants reduce the richness and evenness of marine communities: A review and meta-analysis. *Environ Pollut* 2009, 157(6), 1745–1752.

80. FAO Fisheries and Aquaculture Department. *The State of World Fisheries and Aquaculture.* Food and Agriculture Organization of the United Nations, Rome, 2016.

81. Teh LCL, and Sumaila UR. Contribution of marine fisheries to worldwide employment. *Fish Fish* 2013, 14, 77–88.

82. Trasande L, Zoeller RT, Hass U, Kortenkamp A, Grandjean P, Myers JP, DiGangi J et al. Estimating burden and disease costs of exposure to EDFCs in the EU. *J Clin Endocrinol Metab* 2015, 100(4), 1245–1255.

83. Attina TM, Hauser R, Sathyanarayana S, Hunt PA, Bourguignon JP, Myers JP, DiGangi J, Zoeller RT, and Trasande L. Exposure to endocrine-disrupting chemicals in the USA: A population-based disease burden and cost analysis. *Lancet Diabetes Endocrinol* 2016, 4(12), 996–1003.

84. Gabrys J, Hawkins G, and Michael M (eds). *Accumulation. The Material Politics of Plastic.* Routledge, Abingdon, 2013.

85. Brink P, Schweitzer JP, Watkins E, Smat M, Leslie H, and Galgani F. T20 Task Force Circular Economy: Circular economy measures to keep plastics and their value in the economy, avoid waste and reduce marine litter. G20 Insights. G20 Germany 2017. 2017.

86. EUNOMIA Research & Consulting. *Measures to prevent marine plastic pollution: The trouble with targets and the merits of measures.* EUNOMIA Research & Consulting, Bristol, 2016.

87. Wagner M, Scherer C, Alvarez-Munoz D, Brennholt N, Bourrain X, Buchinger S, Fries E et al. Microplastics in fresh- water ecosystems: What we know and what we need to know. *Environ Sci Eur* 2014, 26(12), 9.

88. PricewaterhouseCoopers AG WPG, Albrecht P, Brodersen J, Horst DW, and Scherf M. Reuse and recycling systems for selected beverage pack- aging from a sustainability perspective. Deutsche Umwelthilfe e.V. & DUH Umweltschutz-Service GmbH. 2011. http://www.duh.de. Accessed December 2016.

89. Thevenon F, Carroll C, and Sousa J (eds). Plastic debris in the ocean: the characterization of marine plastics and their environmental impacts, situation analysis report. IUCN, Gland, 2014, p. 52.

90. Diamanti-Kandarakis E, Bourguignon JP, Giudice LC, Hauser R, Prins GS, Soto AM, Zoeller RT, and Gore AC. Endocrine-disrupting chemicals: An endocrine society scientific statement. *Endocr Rev* 2009, 30(4), 293–342.

Microplastics in Freshwater

CHAPTER **11**

Microplastics in Freshwater

Mohammad Rashid, Shariq Shamsi
and Khwaja Salahuddin Siddiqi

CONTENTS

The natural environment is persistently being exposed to contaminants of emerging concern which does not limit itself to well-known pollutants but also that of microscopic origin [1].

Plastic has for some time been known to be a major component of riverine pollution [2–6], and its degradation products have been noted as a potential issue for soil environments [7]. Indiscriminate disposal places a huge burden on waste management systems, allowing plastic wastes to infiltrate ecosystems, with the potential to contaminate the food chain. Of particular concern is the reported presence of microscopic plastic debris or microplastics (debris ≤1 mm in size) in aquatic, terrestrial and marine habitats [8].The accumulation of small plastic items (<5 mm) in aquatic environments was first reported in marine settings in the 1970s. In 2004, Thompson et al. reported the accumulation of microscopic pieces of plastic and fiber around the United Kingdom. Using archived plankton samples, they showed increases in the abundance of these microscopic particles from the 1960s to the 1990s [9].

By 2050, however, it is anticipated that an extra 33 billion tonnes of plastic will be added to the planet [10]. At present, the increased awareness of the growing production and accumulation of plastic pollution in the environment has brought greater focus to the need for development of policies and management strategies [1].

There is concern that long-term exposure may lead to bioaccumulation of submicron particles with wider implications for environmental health [11–13].

11.1 PLASTICS AND MICROPLASTICS: A HISTORICAL OVERVIEW

The amalgamation of new synthetic chemicals with the engineering capabilities of mass production has made plastics one of the most popular materials in modern times. The discovery of vulcanization of natural rubber by Charles Goodyear [14] led to a number of attempts to develop synthetic polymers including polystyrene (PS) and polyvinyl chloride (PVC). The first synthetic polymer to initiate mass production was Bakelite, a phenol-formaldehyde resin, developed by the Belgian chemist Leo Baekeland in 1909 [15]. Around 1930, the modern forms of PVC, polyethylene terephthalate (PET), polyurethane (PUR) and a more processable PS, were developed [16]. The early 1950s witnessed the development of high-density polyethylene (HDPE) and polypropylene (PP). In the 1960s, the natural resources such as the bacterial fermentation of sugars and lipids were utilized for the development of plastic materials [17] which include polyhydroxyalkanoates (PHA), polylactides (PLA), aliphatic polyesters and polysaccharides [18]. PLA is a commonly used form of bioplastic in end-segment consumer market products these days.

11.2 PLASTICS

The term plastic is used to describe plastic polymers to which various additives are added to give desirable properties to the final product (OECD 2004). Plastics are generally categorized into two types: thermoplastic, which soften on heating can be remolded, and thermosetting, where cross-linking in the polymers occurs and they cannot be re-softened and remolded [8].

The plastic industry is expertly advancing in the arena of nanotechnology innovation. The recent technological advances have introduced the development of newer applications of elements based on nanomaterials that are now producing plastic nanocomposites which include materials that are reinforced with nanofillers for weight reduction, carbon nanotubes (CNTs) for improved mechanical strength and nano-silver as an antimicrobial agent in plastic food packaging materials. However, microplastics have been studied mostly in the context of the marine environment and have been found to be a major constituent of anthropogenic marine debris [8]. By 2020, the share of nanocomposites among plastics in the US will be 7% [19]. It is estimated that only 9% of the 9 billion tons of plastic ever produced has been recycled according to Plastic Recycling Data from UN Environment 2018. If this situation continues, the landfills will have 12 billion tons of plastic waste by 2050. It is therefore suggested that a sustainable waste management and an alternative to plastic materials is needed. In India, 15,000 tons of plastic waste is generated daily, out of which only 9000 tons is collected and recycled; 60% of this plastic is recycled which is higher than the global average of 14%. The remaining plastic material that is not recycled goes to landfills and oceans. It is hazardous to both mammals and marine life, which finally enters the food chain. Every part of the world, even the Arctic ice and mountains are polluted with macro-, micro- and nanoplastic waste material.

11.3 MICROPLASTICS: THE EVOLUTION OF PLASTIC POLLUTION

Apart from plastic litter, MPs (microplastics) have undoubtedly been present in the environment for many years. Concern has been raised that microscopic plastic debris may also be harmful to the environment and to human health [20]. The term microplastic is used to describe the tiny plastic pieces in freshwater systems around the globe including lakes and rivers. In the literature on freshwater habitats, few studies on microplastics existed prior to the 21st century; Hays and Cormons (1974) [21] and Moore et al. (2011) [50] documented plastic particles (<5 mm) in the rivers of North America, while Faure et al. (2015) [54] reported on microplastics in Lake Geneva.

The term "microplastics" commonly refers to plastic particles with a diameter <5 mm. It has been suggested that the term microplastics be redefined as items <1 mm to include only particles in the micrometer size range [22,23], and the term "mesoplastic" introduced to account for items between 1 and 2500 mm [24]. Lambert et al. [25] described macroplastics as >5 mm, mesoplastics as <5 to >1 mm, microplastics as <1 mm to >0.1 μm, and nanoplastics as <0.1 μm.

Generally, MPs are divided into categories of either primary or secondary MPs. Primary MPs are manufactured as such to be used either as resin pellets to produce larger items or directly in cosmetic products such as facial scrubs (known as microbeads; see Figure 11.1) and toothpastes or in abrasive blasting (e.g., to remove lacquers). Compared to this restrained use, secondary MPs are formed from the disintegration of larger plastic debris. They include fragments arising from plastic bottles, bags and other packaging materials [9].

11.4 SOURCES OF PLASTICS AND MICROPLASTICS INTO THE FRESHWATER ENVIRONMENT

There are various ways and sources in which plastic waste enters into the freshwater environment. Land littering is an important environmental and public issue, and measures are now needed to reduce damage to the environment [26].

MPs have different environmental release pathways (Figure 11.2).

- Primary MPs like polyethylene, polypropylene and polystyrene particles in cleaning and cosmetic products enter aquatic systems through household sewage discharge [28–30].

FIGURE 11.1   Microbeads.

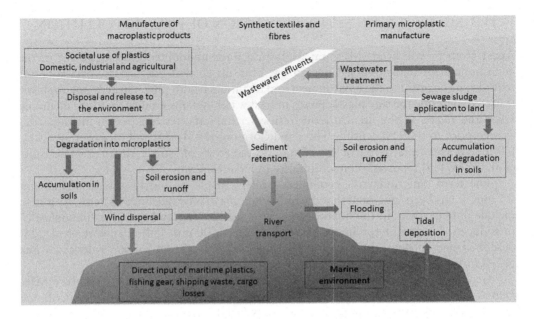

FIGURE 11.2   Conceptual diagram of microplastic sources and flows among anthropogenic, terrestrial, freshwater and marine environmental compartments. (Source: Horton AA, Walton A, Spurgeon DJ, Lahive E, and Svendsen C. Microplastics in freshwater and terrestrial environments: Evaluating the current understanding to identify the knowledge gaps and future research priorities, *Sci Total Environ* 2017, 586, 127–141.)

- Beaches receive primary MPs from spills at inland factories that made their way to coastlines via rivers, streams, and storm-water drains [31]. In recent freshwater literature, large amounts of virgin plastics have been found near plastic production and processing plants along the Rhine River [32] and Danube River [33].
- Secondary microplastics arising from litter such as a discarded drinking-water bottles, fishing gear or plastic film can disintegrate into multiple smaller items over long timescales, and even if all inputs of plastic to aquatic environments were to cease with immediate effect, an increase in microplastic particles would still exist as a result of the fragmentation of large plastic litter already present in the environment.

A 2017 study documented that as a result of slow degradation rates of plastics and the lake's long hydraulic residence time, some microplastics that exist in Lake Erie are likely fragments from some of the very first plastic products that entered the consumer market [34].

- Microplastics that come from washing clothes are mainly polyester, acrylic and polyamide [35,23].

A study reports maximum fiber loads released from washing of clothes as >700,000 fibers from a 6 kg wash of acrylic clothes [36].

- Plastic films used for crop production are considered to be one of the most important sources of plastic contamination of agricultural soils [37–39].
- A wide assortment of microplastics includes particles from plastic mulching, abrasion from car tires or abrasion of synthetic paints [40]. While these plastics

may be defined as secondary source microplastics, they are still different from particles that arise owing to environmental weathering, and so, they can safely be categorized separately.

11.5 OCCURRENCE IN FRESHWATER SYSTEMS

As compared to marine environments, the occurrence of microplastics in freshwater environments is less understood [41,42]. The early investigations of freshwater systems suggest microplastic presence and interactions to be equally as far reaching as observed in marine systems. In the marine environment, microplastics cover a wide range of habitats: at the sea surface, on shorelines, in the deep sea, in sea ice and in biota. They are reported in geographic locations around the globe and widely reported in different depositional environments. They affect the ecological functioning of biogenic habitat like pelagic open ocean habitats to deep-sea benthic habitats and also the coastlines of all continents. They are found even within organisms far from the human presence, such as in deep-sea corals [9].

Identification and quantification of plastic particles are very sensitive and are key steps to address the plastic contamination. The detection and analytical confirmation of purified samples of MPs require access to sophisticated equipment like micro-FTIR [43], Raman spectroscopy [33] scanning electron microscopy, and energy-dispersive X-ray spectrometry (SEM-EDS) (Rocha- Santos and Duarte 2015) to analyze the chemical constituents of microplastics. After pyrolysis, gas chromatography and mass spectrometry (GC-MS) can also be carried out, but the information about shape and size of particle is lost [44].

Recently, substantial data have been collected on microplastics in freshwater habitats by Faure et al. [54]. The presence of primary and secondary microplastics in Lake Geneva, between Switzerland and France, have been found. Moore et al. emphasized that microplastics entered the environment upstream of oceans [50]. The MP concentrations in surface water samples of the Rhine river (Germany) were reported at an average of 892,777 particles/km², with a peak concentration of 3.9 million particles/km² [32]. In river shore sediments, the number of particles ranged from 228 to 3763 and 786 to 1368 particles/kg along the rivers Rhine and Main (Germany), respectively [45]. Su et al. showed microplastic pollution levels during 2015 in Lake Taihu in China and found that microplastics reached $0.01 \times 10^6 - 6.8 \times 10^6$ items/km² in plankton net samples. The microplastics distribution in plankton net samples was different from that in sediments, and microplastics were dominated by fibers 100–1000 µm in size [46]. The Three Gorges Reservoir in China revealed high surface-water contamination with fibers being the most abundant plastics. High-density microplastics were likely to deposit in the sediments [47]. Research on microplastics in freshwater systems is gaining impetus, and microplastics are now recorded in freshwater systems of several continents around the globe.

11.6 FATE AND TRANSPORT IN FRESHWATER SYSTEMS

Microplastics have been detected in rivers and streams around the world. Their measurements shed light on their occurrence in freshwater systems, but they are fragmentary in both space and time [48]. The basis of research on microplastic pollution depends on (a) vertical and horizontal distribution in the aquatic environment and factors that affect the distribution, (b) changes of material properties due to weathering and their consequences on the fate, behavior and transport of MP and (c) MP toxicity and their dependence on different material properties [49].

11.6.1 Transport of MPs

The quantity of microplastics present in the freshwater systems is in direct co-relation with factors like human population density proximal to the water body, proximity to urban centers, proximity to major river inflows, precipitation events, movement of surface water currents, water residence time, size of the water body, the type of waste management used, amount of sewage overflow, and atmospheric depositions [50–55,34]. The population density at or near the freshwater body remains the most influential factor. Eriksen et al. [52] reported that Lake Erie had microplastic particle counts (466,305 particles/km^2) that were 70 and 38 times greater as compared to the less populated Lakes Huron (6541 particles/km^2) and Superior (12,645 particles/km^2). Even within Lake Erie, sites near cities had more microplastics, including more pellets compared with rural shorelines [56]. Also, the quantity of plastic litter which is released to the environment is directly proportional to the international differences in societal attitudes, education and investment in waste management infrastructure. In 2010, China was unable to manage its 76% of plastic waste (8.82 million metric tonnes) as opposed to 2% (0.28 million 466 metric tonnes) of the United States [57].

Transport of plastic particles within the river systems is largely affected by the same factors affecting sediment transport, such as hydrological characteristics and environmental conditions [103]. Change in river depth or velocity (e.g., on a bend) may lead to deposition of particulate matter, whereas high velocity flood conditions and erosion could result in mobilization of previously sedimented particles, in addition to the introduction of particles via runoff [104,105]. Surrounding land use can also affect the dynamics of sediment and particulate transport within a river due to erosion, use of soils, irrigation and runoff [106]. Presence of plastic in riverine systems may also be subjected to *in situ* degradation, either by light or mechanical fragmentation [107].

11.6.2 Environmental Persistence and Degradation

Plastic can undergo photo degradation, thermal degradation, mechanical degradation, thermo-oxidative degradation and hydrolysis [22]. All these degradation processes of plastics under environmentally relevant conditions are affected by almost all of the types of degradation mechanisms. For HDPE (high-density polyethylene) and nylon, degradation is primarily via UV-B photooxidation followed by thermo-oxidation. Degraded products of micro- or potentially even nano-sized dimensions [108,109] may further be degraded (e.g., biological) where carbon in the matrix is converted to carbon dioxide and incorporated into biomass [22]. However, UV degradation of plastics floating in water is delayed by lower temperatures and oxygen levels relative to on land, making conversion of macroplastics to microplastics much more rapid on beaches than in the water [22,58,110].

The plastic degradation processes carried out in the laboratory focuses on a single mechanism such as photo, thermal or biodegradation. The time scaling of the process from the abandonment of a waste, its arrival and persistence in the aquatic environment is not known. This scenario is accompanied by a physical fragmentation into particles of increasingly smaller sizes and also by a chemical functionalization due to the photooxidation of the macromolecular chains [60] (Table 11.2).

Even though all the degradation mechanisms work together, the studies discuss weight loss, changes in tensile strength, breakdown of molecular structure and identification of specific microbial strains to utilize specific polymer types [61]. The degradation processes

TABLE 11.1   Crystallinity of Polymers

Polymer Type	Crystallinity
Natural rubber	Low
Polyethylene–low density	45%–60%
Polyethylene–high density	70%–95%
Polypropylene	50%–80%
Polystyrene	Low
Polycarbonate	Low
Polyvinyl chloride	High
Polylactic acid	37%
Polyethylene terephthalate	Described as high in and as 30%–40% in [59]

TABLE 11.2   Weathering and Degradation of Microplastics

Variables Influencing the Weathering of MP	Characteristics That Affect Degradation
a. Polymer type b. Polymer properties such as density and crystallinity c. Embedded additives in the polymer structure d. Environmental exposure conditions e. Type of weathering processes that can occur even without UV light [49]	a. Complex structure: They are not easily biodegradable. b. Crystallinity: The crystalline region is more ordered with tightly structured polymer chains. Crystallinity affects physical properties (like density and permeability), which in turn affects their hydration and swelling behavior. This affects availability of sorption sites for microorganisms (Table 11.1). c. Presence of stabilizers: Stabilizers (antioxidants and antimicrobial agents) act to prolong the life of plastics. d. Biological ingredients act to decompose the plastic in shorter time frames [21].

are defined for the degradation mechanism under investigation (e.g., thermal degradation) and the result is generated. In contrast, particle formation rates are often not investigated. This is important because polymers such as PE do not readily depolymerize and generally decompose into smaller fragments and further disintegrate into increasingly smaller fragments, eventually forming nanoplastics [62].

For the purpose of studying breakup events to describe the distribution of fragment sizes, the kinetic models are introduced. These models simulate the transport and fate of buoyant and nonbuoyant plastic debris (MPs) in freshwater systems. These models work on the rate equations, which assume that each particle is exposed to an average environment, mass is the unit used to characterize a particle and the size distribution is spatially uniform [63,64]. The models generally focus on chain scission in the polymer backbone through (a) random chain scission (all bonds break with equal probability) characterized by oxidative reactions; (b) scission at the chain midpoint dominated by mechanical degradation; (c) chain-end scission, a monomer-yielding depolymerization reaction found in thermal and photodecomposition processes and (d) in terms of in homogeneity (different bonds have different breaking probability and dispersed throughout the system) [65–67]. The estimation of degradation half-lives has also been measured for strongly hydrolysable polymers through the use of exponential decay equations [61,68,69], though the studies need further investigation in case of chemically resistant plastics [69].

Henceforth, the environmental degradation processes encapsulate MP fragmentation into increasingly smaller particles including nanoplastics, chemical transformation of the plastic fragments, degradation of the plastic fragments into nonpolymer organic molecules and the transformation/degradation of these nonpolymer molecules into other compounds [61].

11.6.3 Interaction with Other Compounds

The exchange of additives or contaminants between microplastic particles and the surrounding water depends on

- Concentration gradient,
- Matrix in which the microplastics are present,
- Physical and chemical properties of the plastic polymer and
- Degradation processes acting upon microplastic particles [111,112].

The sorption of hydrophobic pollutants to the noncrystalline regions of plastic polymers occurs, and the smaller additives tend to move out of plastics fast. Apart from the compounds used in plastic manufacturing, there is also potential for exposure of aquatic biota to other contaminants that adsorb to microplastic particles. Microplastics are typically hydrophobic and have large surface areas, allowing them to accumulate organic pollutants such as polycyclic aromatic hydrocarbons (PAHs), polybrominated diphenylethers (PBDEs), polychlorinated biphenyls (PCBs) and dichlorodiphenyltrichloroethane. Ashton et al. [70] reported concentrations of metals in composite plastic pellet samples retrieved from the high-tide line along a stretch of coastline in southwest England. In the freshwater environment, MPs are likely to co-occur with other contaminants such as pharmaceuticals, personal care products, flame retardants and other industrial chemicals which enter the environment as parts of complex solid and liquid waste streams [21].

The sorption processes takes place through physical and chemical adsorption as well as pore-filling processes. Physical adsorption is the reversible sorption to surfaces of the polymer matrix, that is, it does not involve the formation of covalent bonds. Chemical adsorption involves chemical reactions between the polymer surface and the sorbate which generates new chemical bonds at the polymer surface depending on how aged the polymer surface is. The changes in pH, temperature and ionic strength of the localized environment influence these processes [71]. In pore-filling processes, the hydrophobic pollutants with lower molecular weights will more easily move through a polymer matrix with larger pores.

11.6.3.1 Factors Influencing Adsorption Process

- *Particle size and texture*: Decreased particle size increases the potential for surface chemical interactions and, thus, binding with hydrophobic chemicals. Also, the physically weathered particles have more surface area and may be exposed to high levels of UV radiation and wind.
- *Structure of the polymer*: Polymers that have structures with short and repeating units, a high symmetry and strong interchain hydrogen bonding will have a lower sorption potential.
- *Polymer density and crystallinity*: The sorption and diffusion of hydrophobic contaminants are most likely to take place in the amorphous area of a plastic

material, because the crystalline region consists of more ordered and tightly structured polymer chains.

- *Pollutants present*: Pollutants with lower molecular weight have high sorption potential.
- *Surrounding environment* [27].

11.7 EFFECTS OF PLASTICS AND MICROPLASTICS ON FRESHWATER ECOSYSTEMS

In aquatic ecosystems, the mobility and degradation of plastics potentially generate a mixture of parent materials, fragmented particles of different sizes, and other nonpolymer degradation products. Such complex mixture of plastics and associated chemicals that change in time and space influence the biota significantly [21].

The first interaction between the biota and microplastics is biofouling, which involves colonization of the particle surface by a biofilm [113]. It may lead to transport of biota and/ or invasive species [113,114]. Another early interaction is ingestion, for example, oysters and mussels. [115,116]. Other possible interactions could be stress responses such as inflammation or oxidative stress at tissue and cellular levels [117,118]. Microplastics may also serve as a medium to expose biota to environmental contaminants adsorbed to their surfaces [119].

11.7.1 Ingestion of MPs and its Biological Impact

MPs may be taken up from the water column and sediment by a range of organisms directly through ingestion or dermal uptake most importantly through respiratory surfaces (gills) [21].

One example is the zebrafish (*Danio rerio*) in which PS particles accumulate in the gills (5 and 20 μm), gut (5 and 20 μm) and liver (5 μm) [72].

Among the clearest evidence to date that freshwater species are exposed directly to microplastic pollutants is provided by the work of Windsor et al. [73], who recently documented the presence of microplastics in the guts or other tissues of several species of invertebrates (including insect larvae) (Figure 11.3) collected at locations along the rivers Usk, Taff and Wye in South Wales [73].

Although plastic is largely excreted following ingestion, evidences suggest that microplastics can be retained in the gut over timescales beyond those expected for other ingested matter [74]. Particles may even cross the gut wall and translocate to other body tissues, with unknown consequences. [11,74]. Some species are capable of rapid excretion while others accumulate and/or mobilize microplastics into their circulation. Sanchez et al. [75] investigated gudgeon (*Gobio gobio*) caught in 11 French streams and found MP in the digestive tract of 12% of the fish. However, the rate of MP ingestion in different species of fishes certainly depends on their feeding strategy. Rosenkranz et al. [76] demonstrated that the water flea *Daphnia magna* rapidly ingested MP under laboratory conditions. MP (0.02 and 1 mm) appear to cross the gut epithelium and accumulate in lipid storage droplets inducing severe effects. Imhof et al. [77] reported the uptake of MP by annelids (*Lumbriculus variegatus*), crustaceans (*D. magna* and *Gammarus pulex*), ostracods (*Notodromas monacha*) and gastropods (*Potamopyrgus antipodarum*). Karami et al. observed histological alterations in the gills and blood chemistry parameters (such as plasma cholesterol levels, blood HDL levels) of African catfish (*Clarias gariepinus*) at low

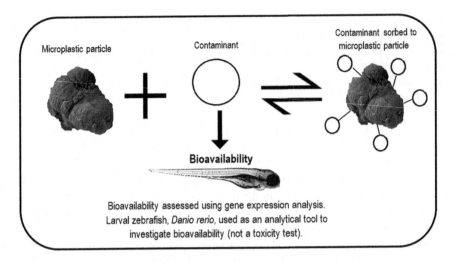

FIGURE 11.3   Caddisfly larva with incorporated plastic.

concentrations of HDPE fragments (50 μg/L) and severe changes (like epithelial sloughing, hyperplasia, extensive cell sloughing) at a higher particle concentration (500 μg/L) [78].

Surprisingly, many species are able to identify particles with nutritional value. Studies with labeled bacteria have shown that some ciliates (estuarine oligotrichs) and flagellates prefer bacteria over MPs, while other species (estuarine scuticociliates; e.g., *Uronema narina*) cannot discriminate between bacteria and MPs. Therefore, we hypothesize a similar pattern regarding species-specific size and taste discrimination: some species have the tendency of directly feeding on available MPs in the size range of their food, while more selective feeders avoid MP ingestion [72].

The ingestion of MPs may lead to the development of physical stress in organisms. The extent of physical stress depends on:

- *Particle size*: Larger particles are harder to digest.
- *Particle shape*: Needle-like particles attach more readily to internal and external surfaces.
- *Physical irritation*: Smaller and angular particles are more difficult to dislodge than smooth spherical particles and therefore may cause blockage of gills and the digestive tract.
- *Exposure to secondary MPs*: MPS of mean particle size 2.6 μm may cause elevated mortality, increased interbrood period and decreased reproduction at very high MP levels [21].

Research to date has revealed the ingestion of microplastics in a wide range of species at many organizational levels with different feeding strategies, including filter feeders and predators. Apart from accumulation of particles in organisms at lower trophic levels [74], there is also evidence for the trophic transfer of microplastic particles, such as from mussels to crabs [79]. The potential of MP transfer from meso- to macro-zooplankton using PS microspheres (10 μm) at much lower concentrations of 1000, 2000 and 10,000 particles/mL was demonstrated [28].

Because excretion rates are unknown and MP uptake is often defined as particles present in the digestive tract (i.e., the outside and not the tissues of an organism), it is

FIGURE 11.4   Bioavailability.

not yet established whether the trophic transfer of MP also results in a bioaccumulation or biomagnification. However, MPs are transferred from the prey to the apex predator, where, in certain situations, they can be retained for longer periods in the body of the latter [12]. The functioning of the ecosystem is directly influenced by MP physical and chemical characteristics, biological aspects like molecular targets, the bioavailability of the MPs and the penetration of submicron MPs into the cells (Figure 11.4).

11.7.2 Chemical Effects

The plastic materials are highly functional compounds of synthetic polymer and additives like plasticizers, flame retardants or colorants). The monomers, catalysts, stabilizers and additives leach to produce chemical toxicity [72].

Many of the chemicals associated with plastics have been identified as either toxic or endocrine disruptors, including bisphenol-A, phthalates such as di-n-butyl phthalate and di-(2-ethylhexyl) phthalate, polybrominated diphenyl ethers (PBDEs) and metals used as coloring agents [80,81,102,109]. These additives are weakly bound or not bound at all to the polymer molecule, and as such, these chemicals will leach out of the plastic over time. Such releases can be facilitated in environments where particle dispersal is limited and so the plastics will experience UV degradation and high temperatures [22]. The locations where microplastics may accumulate in soil and surface waters are then likely to be subjected to the possible release of these chemicals from plastics and are transferred to water, sediment and organisms. Locations closer to the source of plastic litter MPs are likely to have higher concentration of plastic additives. Lithner et al. [81] showed that different plastic items can leach toxic chemicals into water that can cause varying effects on *Daphnia magna* [27].

For instance, with a liquid-to-solid (L/S) ratio of 10 and 24 h leaching time, leachates from polyvinyl chloride (PVC), polyurethane (PUR) and polycarbonate (PC) were found to be most toxic with EC50 values of 5–69 g plastic/L[81]. Higher L/S ratios and longer leaching times resulted in leachates from plasticized PVC. The epoxy resin products were the most toxic at EC50 of 2–235 g plastic/L. There are many challenges associated with the characterization of leachates owing to the potential diversity of physicochemical properties of chemical breakdown products [21].

Several laboratory studies depict the capacity of MPs to modify adverse effects of chemicals by affecting the bioavailability or acting as an additional stressor. Besseling et al. [82] observed a decreased bioaccumulation of polychlorinated biphenyls in lugworms at higher doses of PS particles. Karami et al. [78] modulate the impacts of phenanthrene on biomarker responses in African catfish (*Clarias gariepinus*). Paul-Pont et al. [83] detected the modulations of adverse effects by an exposure to phenanthrene-loaded low-density polyethylene (LDPE) fragments (African catfish) and PS beads and fluoranthene (*Mytilus* spp.), respectively. However, Gouin et al. [84] and Koelmans et al. [85] highlighted the minor influence of MPs as vectors for the bioaccumulation of pollutants considering they are outcompeted by naturally occurring matter.

In addition to the potential of MPs in influencing the bioavailability of toxic compounds, Besseling et al. [86] suggested that MPs can interfere with intra- and interspecies signaling (e.g., pheromones and kairomones) as an integral component of aquatic biocoenosis regulating predator–prey interactions as well as population and community structures [87]. It can lead to maladaptive responses in both the signaler and receiver [88]. But the fact still remains unclear whether MPs are the culprit for maladaptive responses since there exists an abundance of additional particulate organic and inorganic matter in aquatic ecosystems as well.

11.7.3 Biofilm-Related Impact

Colonization of MPs with microbes and adsorption of biopolymers increase the nutritional value and improve the "taste," making them more attractive for biota in contrast to the colonization of MPs with pathogens [89] and toxic algae/bacteria, which might induce chemical toxicity or avoidance of "bad tasting" MPs.

11.7.3.1 Effect of Biofilm
- Biofilm formation increases the density of floating MPs and leads to sedimentation of these low-density particles [90].
- Formation of hetero-aggregates consist of particulate matter (MPs as well as other suspended solids) and microbes (e.g., protozoans, algae) with biopolymers acting as binders. Due to the abundance of microscopic particles, the availability to micro-feeders (e.g., protozoans, planktonic crustaceans) decreases, and large hetero-aggregates are accessible to macro-feeders (e.g., planktivorous fishes).

11.8 BIOLOGICAL EFFECTS OF SUB-MICROMETER PLASTICS

Nano- and microplastics often come from unintentional anthropogenic rather than engineered processes. They are intended to be biologically active, even when intentionally produced. For both engineered nanomaterials and nano- and microplastics, it is therefore useful to consider their intended use and properties when evaluating their potential environmental risk [91].

The smaller-sized polymer particles, with a larger fraction of the molecules, will be present on the surface of the particle, which can lead to an increase in chemical reactions and biological interactions. For example, smaller particles (on a mass basis) have a larger adsorption capacity as compared to larger particles [92], resulting in relevant vector effects. Also, they have potential to cross biological barriers. Engineered nanoparticles (ENPs) are able to cross cell membranes and become internalized, although the uptake of ENPs is size dependent, with uptake occurring by endocytosis or phagocytosis [93]. They are stored inside vesicles and mitochondria [104]. Cellular responses include oxidative stress, antioxidant activity and cytotoxicity [94].

The leaching of molecules from particles and the transformation processes, such as oxidation/reduction, interaction with macromolecules, light exposure and biological transformation, can significantly influence the integrity, behavior and persistence of nanomaterials in aquatic media [95].

Desai et al. [96] showed that polylactic (polyglycolic) acid nanoparticles of 100 nm copolymer had a tenfold higher intracellular uptake in an *in vitro* cell culture when compared to 10 μm particles made of the same material. Several cytotoxic, genotoxic, inflammatory and oxidative-stress responses in mammalian and fish systems have been reported. Therefore, there is a need to understand the molecular and cellular pathways and the kinetics of absorption, distribution, metabolism and excretion mechanisms that may be unique to MPs in the nano-size range. Aging is currently not incorporated in standard ecotoxicity test protocols, but has been proposed for engineered nanomaterials [97].

11.9 FRESHWATER MICROPLASTICS: CHALLENGES FOR REGULATION AND MANAGEMENT

"Microplastics" refer to a large group of polymers with various chemical and physical properties, originating from different sources and entering the environment via different pathways [21,98]). Verschoor [99] identified five commonly applied criteria to define MP: (1) synthetic materials with high polymer content, (2) solid particles, (3) <5 mm, (4) insoluble in water and (5) not degradable. However, discussion goes on whether tire abrasion should be considered as "microplastics" and the same applies to the definition of a lower limit for particle size. The smaller the particles are, the more species might potentially ingest those. Furthermore, smaller particles can permeate through membranes and, hence, pose a higher risk for adverse effects in organisms. Thus, globally agreed standardization approaches covering the whole range from sampling to effect assessment are required in order to provide a basis for risk assessment and regulatory options.

As a first approach for MP assessment in freshwater environments, Miklos et al. [100] suggested a modular system initiating with the quantification of selected indicator polymers. When the concentration of these polymers exceeds a certain level, more specific analyses should be conducted. These subsequent analyses can take various criteria (such as polymer type, size, shape, additives, etc.) into consideration to further categorize the particles and support the selection of adequate mitigation measures. Here, approaches from chemical regulation might serve as examples where chemicals can be categorized based on molecular similarities (e.g., PAHs, PCBs, etc.), by field application (e.g., pesticides), or according to their mode of action (e.g., endocrine disruptors). Similarly, MPs can also be grouped together based on their physicochemical properties (e.g., polymer type, density), by their application (e.g., cosmetics, carrier bags, electrical devices) or by (eco)toxicological impacts.

The chemical effects could be the result of polymers themselves or their additives, or a combination of both, while mechanical effect could depend on particle size, shape or a combination of both. Since the (eco)toxicologists face the challenge to test the effects of myriads combination, there are efforts to prioritize and start with the presumed most harmful combinations. Ideally, these results will be transferable to a group of similar combinations.

The physicochemical properties of MP might lead to different behavior in test systems. High-density polymers will settle as sediments whilst the low-density polymers float on the surface. They are available only for surface-feeding organisms and are at higher risk in environmental systems as they might feed selectively on floating materials and accumulate them from the water phase. Chemical testing is usually not focused on this feeding type. Above all, scientists should contemplate that organisms are adapted to natural particles of different materials (sand, clay or similar), but with similar properties as MP, in their natural habitats. It is critical to perform tests on MP particles in comparison with such natural particles. This implicates especially to tests regarding the "Trojan horse effect"—the transport of hydrophobic substances via MP into organisms [72]. Studies need to address whether or not there are differences in the sorption of chemicals to MP versus natural particles.

To reduce the influx of MPs into the aquatic environment, there is a need to improve the management of wastewater and solid waste apart from affecting the production or application of plastics. Starting with product design, we can apply a range of possibilities that includes degradable polymers, polymers with high recycling quotas or a product design promoting a long and circular product life to reduce waste [101].

One of the most sustainable measures would be a social change, and to achieve such long-term objectives, policy can apply so-called persuasive instruments such as public information, environmental education and funding of research and development. Precisely because we currently know little about the consequences of MPs in aquatic systems, we should develop and implement measures to reduce further emissions.

11.10 CONCLUSION

The distributions and abundance of microplastics in freshwater and marine systems are likely to increase with the growing input of plastic into the environment unless the production and use of polymeric materials is legally forbidden. We are inadvertently consuming micro- and nanoparticles of all types of plastic and polymeric materials through the food chain. Freshwaters may be similar to marine systems wherein studies have shown that various types of biota, from microorganisms to vertebrates, across numerous trophic levels, interact with microplastics, with some negative impacts documented in laboratory studies. Further work to investigate potential impacts of microplastics on freshwater biota at environmentally realistic concentrations of microplastic is needed to reach any firm conclusion.

REFERENCES

1. Jeftic L, Sheavly SB, and Adler E. Marine litter: A global challenge. In: Meith N (ed) *Regional Seas*. United Nations Environment Programme, 2009.
2. Waters CN, Zalasiewicz J, Summerhayes C, Barnosky AD, Poirier C, Galuszka A, Cearreta A et al. The Anthropocene is functionally and stratigraphically distinct from the Holocene. *Science* 2016, 351(6269).
3. Balas CE, Williams AT, Simmons SL, and Ergin A. A statistical riverine litter propagation model. *Mar Pollut Bull* 2001, 42(11), 1169–1176.
4. Williams AT, and Simmons SL. Movement patterns of riverine litter. *Water Air Soil Pollut* 1997, 98(1), 119–139.
5. Williams AT, and Simmons SL. The degradation of plastic litter in rivers: Implications for beaches. *J Coast Conserv* 1996, 2(1), 63–72. doi:10.1007/bf02743038

6. Tudor DT, and Williams AT. Development of a 'matrix scoring technique' to determine litter sources at a Bristol channel beach. *J Coast Conserv* 2004, 10(1–2), 119–127.

7. Klemchuk PP. Degradable plastics – a critical-review. *Polym Degrad Stab* 1990, 27(2), 183–202.

8. Galloway TS. Micro- and Nano-plastics and Human Health. In: Bergmann M, Gutow L, and Klages M (eds) *Marine Anthropogenic Litter*. Springer, Cham, 2015.

9. Medrano DE, and Richard Thompson R. *Aberdeen, handbook on Occurrence, Fate, and Effect of Microplastics in Freshwater Systems.*

10. Browne MA, Crump P, Niven SJ, Teuten E, Tonkin A, Galloway T, and Thompson R. Accumulation of microplastic on shorelines worldwide: Sources and sinks. *Environ Sci Technol* 2011, 45(21), 9175–9179.

11. Farrell P, and Nelson K. Trophic level transfer of microplastic: Mytilus edulis (L.) to Carcinus maenas (L.). *Environ Pollut* 2013, 177, 1–3.

12. Setälä O, Fleming-Lehtinen V, and Lehtiniemi M. Ingestion and transfer of microplastics in the planktonic food web. *Environ Pollut* 2014, 185, 77–83.

13. Murray F, and Cowie PR. Plastic contamination in the decapod crustacean Nephrops norvegicus (Linnaeus, 1758). *Mar Pollut Bull* 2011, 62(6), 1207–1217.

14. Stevenson K, Stallwood B, and Hart AG. Tire rubber recycling and bioremediation: A review. *Biorem J* 2008, 12(1), 1–11.

15. Vlachopoulos J, and Strutt D. Polymer processing. *Mater Sci Technol* 2003, 19(9), 1161–1169.

16. Brandsch J, and Piringer O. Characteristics of plastic materials. In: Piringer OG, and Baner AL (eds.) *Plastic Packaging*. Wiley-VCH Verlag GmbH & Co. KGaA, Weinheim, 2008, pp. 15–61.

17. Lambert S. Biopolymers and their application as biodegradable plastics. In: Kalia CV (ed) *Microbial Factories: Biodiversity, Biopolymers, Bioactive Molecules*, vol 2. Springer India, New Delhi, 2015, pp. 1–9.

18. Reddy CSK, Ghai R, and Rashmi KVC. Polyhydroxyalkanoates: An overview. *Bioresour Technol* 2003, 87(2), 137–146.

19. Roes L, Patel MK, Worrell E, and Ludwig C. Preliminary evaluation of risks related to waste incineration of polymer nanocomposites. *Sci Total Environ* 2012, 417, 76–86.

20. Thompson RC, Olsen Y, Mitchell RP, Davis A, Rowland SJ, John AW, McGonigle D, and Russell, AE. Lost at sea: Where is all the plastic? *Science* 2004, 304(5672), 838.

21. Hays H, and Commons G. Plastic particles found in tern pellets, on coastal beaches and at factory sites. *Mar Pollut Bull* 1974, 5(3), 44–46.

22. Andrady AL. Microplastics in the marine environment. *Mar Pollut Bull* 2011, 62(8), 1596–1605.

23. Browne MA, Crump P, Niven SJ, Teuten E, Tonkin A, Galloway T, and Thompson R. Accumulation of microplastic on shorelines woldwide: Sources and sinks. *Environ Sci Technol* 2011, 45(21), 9175–9179.

24. GESAMP. Sources, fate and effects of microplastics in the marine environment: a global assessment. Joint Group of Experts on the Scientific Aspects of Marine Environmental Protection Reports and studies 2015, 90.

25. Lambert S, Sinclair CJ, and Boxall ABA. Occurrence, degradation and effects of polymerbased materials in the environment. *Rev Environ Contam Toxicol* 2014, 227, 1–53.

26. Cierjacks A, Behr F, and Kowarik I. Operational performance indicators for litter management at festivals in semi-natural landscapes. *Ecol Indic* 2012, 13(1), 328–337.

27. Horton AA, Walton A, Spurgeon DJ, Lahive E, and Svendsen C. Microplastics in freshwater and terrestrial environments: Evaluating the current understanding to identify the knowledge gaps and future research priorities, *Sci Total Environ* 2017, 586, 127–141.

28. Zitko V, and Hanlon M. Another source of pollution by plastics: Skin cleaners with plastic scrubbers. *Mar. Pollut Bull* 1991, 22(1), 41–42.

29. Gregory MR. Plastic "scrubbers" in hand cleansers: A further (and minor) source for marine pollutionidentified. *Mar Pollut Bull* 1996, 32(12), 867–871.

30. Fendall LS, and Sewell MA. Contributing to marine pollution by washing your face: Microplastics in facial cleansers. *Mar Pollut Bull* 2009, 58(8), 1225–1228.

31. Gregory MR. Accumulation and distribution of virgin plastic granules on New Zealand beaches. *N Z J Mar Freshw Res* 1978, 12(4), 399–414.

32. Mani T, Hauk A, Walter U, and Burkhardt-Holm P. Microplastics profile along the Rhine River. *Sci Rep* 2015, 5, 17988.

33. Lechner A, Keckeis H, Lumesberger-Loisl F, Zens B, Krusch R, Tritthart M, Glas, M et al. The Danube so colourful: A potpourri of plastic litter outnumbers fish larvae in Europe's second largest river. *Environ Pollut* 2014, 188, 177–181.

34. Cable RN, Beletsky D, Beletsky R, Wigginton K, Locke BW, and Duhaime MB. Distribution and modeled transport of plastic pollution in the Great Lakes, the world's largest freshwater resource. *Front Environ Sci* 2017, 5(45).

35. Habib D, Locke DC, and Cannone LJ. Synthetic fibers as indicators of municipal sewage sludge, sludge products, and sewage treatment plant effluents. *Water Air Soil Pollut* 1998, 103(1), 1–8.

36. Napper IE, and Thompson RC. Release of synthetic microplastic plastic fibres from domestic washing machines: Effects of fabric type and washing conditions. *Mar Pollut Bull* 2016, 112(1), 39–45.

37. Xu G, Wang QH, Gu QB, Cao YZ, Du XM, and Li FS. Contamination characteristics and degradation behavior of low-density polyethylene film residues in typical farmland soils of China. *J Environ Sci Health Part B* 2006a, 41(2), 189–199.

38. Brodhagen M, Peyron M, Miles C, and Inglis DA. Biodegradable plastic agricultural mulches and key features of microbial degradation. *Appl Microbiol Biotechnol* 2015, 99(3), 1039–1056.

39. Kyrikou I, and Briassoulis D. Biodegradation of agricultural plastic films: A critical review. *J Polym Environ* 2007, 15(2), 125–150.

40. Duis K, and Coors A. Microplastics in the aquatic and terrestrial environment: Sources (with a specific focus on personal care products), fate and effects. *Environ Sci Eur* 2016, 28(1,2).

41. Wagner M, Scherer C, Alvarez-Muñoz D, Brennholt N, Bourrain X, Buchinger S, Fries E et al. Microplastics in freshwater ecosystems: What we know and what we need to know. *Environ Sci Eur* 2014, 26(1), 12.

42. Eerkes-Medrano D, Thompson RC, and Aldridge DC. Microplastics in freshwater systems: A review of the emerging threats, identification of knowledge gaps and prioritisation of research needs. *Water Res* 2015, 75, 63–82.

43. Hidalgo-Ruz, V, Gutow L, Thompson RC, and Thiel M. Microplastics in the marine environment: A review of the methods used for identification and quantification. *Environ Sci Technol* 2012, 46(6), 3060–3075.

44. Shafiq M, Qadir A, and Hussain CM. Microplastics as Contaminant in Freshwater Ecosystem: A Modern Environmental Issue. In: Hussain CM (ed.) *Handbook of Environmental Materials Management.*

45. Klein S, Worch E, and Knepper TP. Occurrence and spatial distribution of microplastics in river shore sediments of the Rhine-main area in Germany. 2015.

46. Su L, Xue Y, Li L, Yang D, Kolandhasamy P, Li D, Shi H. Microplastics in Taihu lake. *Environ Pollut* 2016, 216, 711–719.

47. Di M, and Wang J. Microplastics in surface waters and sediments of the Three Gorges Reservoir, China. *Sci Total Environ* 2018, 616–617, 1620–1627.

48. Kooi M, Wezel AV, and Koelmans A. *From the Sewer to the Sea - Microplastic Transport and Fate in Freshwater Systems MICRO 2018, Fate and Impact of Microplastics: Knowledge, Actions and Solutions.* Lanzarote, November 2018, 19–23.

49. Potthoff A, and Oelschlagel K. *The Relevance of Particle Characterization for Microplastic Particles (Mp) in the Environment MICRO 2018, Fate and Impact of Microplastics: Knowledge, Actions and Solutions.* Lanzarote, November 2018, 19–23.

50. Moore CJ, Lattin GL, and Zellers AF. Quantity and type of plastic debris flowing from two urban rivers to coastal waters and beaches of Southern California. *J Integr Coast Zone Manag* 2011, 11(1), 65–73.

51. Zbyszewski M, and Corcoran PL. Distribution and degradation of fresh water plastic particles along the beaches of Lake Huron, Canada. *Water Air Soil Pollut* 2011, 220(1–4), 365–372.

52. Eriksen M, Mason S, Wilson S, Box C, Zellers A, Edwards W, Farley H et al. Microplastic pollution in the surface waters of the Laurentian Great Lakes. *Mar Pollut Bull* 2013, 77(1), 177–182.

53. Free CM, Jensen OP, Mason SA, Eriksen M, Williamson NJ, and Boldgiv B. High-levels of microplastic pollution in a large, remote, mountain lake. *Mar Pollut Bull* 2014, 85(1), 156–163.

54. Faure F, Demars C, Wieser O, Kunz M, and De Alencastro LF. Plastic pollution in Swiss surface waters: Nature and concentrations, interaction with pollutants. *Environ Chem* 2015, 12(5), 582–591.

55. Dris R, Gasperi J, Saad M, Mirande C, and Tassin B. Synthetic fibers in atmospheric fallout: A source of microplastics in the environment? *Mar Pollut Bull* 2016, 104(1), 290–293.

56. Zbyszewski M, Corcoran PL, and Hockin A. Comparison of the distribution and degradation of plastic debris along shorelines of the Great Lakes, North America. *J Great Lakes Res.* 2014, 40(2), 288–299.

57. Jambeck JR, Geyer R, Wilcox C, Siegler TR, Perryman M, Andrady A, Narayan R et al. Plastic waste inputs from land into the ocean. *Science* 2015, 347(6223), 768–771.

58. Anderson JC, Park BJ, and Palace VP. Microplastics in aquatic environments: Implications for Canadian ecosystems. *Environ Pollut* 2016, 218, 269–280.

59. Ehrenstein GW. *Polymeric Materials: Structure, Properties, Applications.* Hanser Gardner Publications, Inc., München, 2012.

60. Askanian H, Delorjestin F, Koumba GB, and Verney V. *Phothochemical Fragmentation of Freshwater (Micro)Plastics Under UV Irradiations MICRO 2018, Fate and Impact of Microplastics: Knowledge, Actions and Solutions.* Lanzarote, November 2018, 19–23.

61. Lambert S, Sinclair CJ, Bradley EL, and Boxall ABA. Effects of environmental conditions on latex dergadation in aquatic systems. *Sci Total Environ* 2013, 447, 225–234.

62. Gigault J, Pedrono B, Maxit B, and Ter Halle A. Marine plastic litter: The unanalyzed nanofraction. *Environ Sci Nano* 2016, 3(2), 346–350. doi:10.1039/c6en00008h

63. Cheng Z, and Redner S. Kinetics of fragmentation. *J Phys A Math Gen* 1990, 23(7), 1233.

64. Redner S. Statistical theory of fragmentation. In: Charmet JC, Roux S, and Guyon E (eds) *Disorder and Fracture*. NATO ASI series, vol 204. Springer, New York, 1990, pp. 31–48.

65. Ziff RM, and McGrady ED. The kinetics of cluster fragmentation and depolymerisation. *J Phys A Math Gen* 1985, 18(15), 3027.

66. McCoy BJ, and Madras G. Degradation kinetics of polymers in solution: Dynamics of molecular weight distributions. *AIChE J* 1997, 43(3), 802–810. doi:10.1002/aic.690430325

67. Rabek JB. Oxidative degradation of polymers. In: Bamford CH, and Tipper CFH (eds) *Comprehensive Chemical Kinetics: Degradation of Polymers*. vol 14. Elsevier Scientific Publishing Company, Amsterdam, 1975.

68. Amulya K, Jukuri S, and Venkata Mohan S. Sustainable multistage process for enhanced productivity of bioplastics from waste remediation through aerobic dynamic feeding strategy: Process integration for up-scaling. *Bioresour Technol* 2015, 188, 231–239.

69. Bartsev SI, and Gitelson JI. A mathematical model of the global processes of plastic degradation in the World Ocean with account for the surface temperature distribution. *Dokl Earth Sci* 2016, 466(2), 153–1156.

70. Ashton K, Holmes L, and Turner A. Association of metals with plastic production pellets in the marine environment. *Mar Pollut Bull* 2010, 60(11), 2050–2055.

71. Delle Site A. Factors affecting sorption of organic compounds in natural sorbent/water systems and sorption coefficients for selected pollutants. A review. *J Phys Chem Ref Data* 2001, 30(1), 187–439.

72. Scherer C, Weber A, Lambert S, and Wagner M. *Interactions of Microplastics with Freshwater Biota, the Handbook of Environmental Chemistry*, Vol 58, pp. 153–180.

73. Windsor FM, Tilley RM, Tyler CR, and Ormerod SJ. Microplastic ingestion by riverine macroinvertebrates. *Sci Total Environ* 2019, 646(1), 68–74.

74. Browne MA, Dissanayake A, Galloway TS, Lowe DM, and Thompson RC. Ingested microscopic plastic translocates to the circulatory system of the mussel, Mytilus edulis (L.). *Environ Sci Technol* 2008, 42(13), 5026–5031.

75. Sanchez W, Bender C, and Porcher JM. Wild gudgeons (Gobio gobio) from French rivers are contaminated by microplastics: Preliminary study and first evidence. *Environ Res* 2014, 128, 98–100.

76. Rosenkranz P, Chaudhry Q, Stone V, and Fernandes TF. A comparison of nanoparticle and fine particle uptake by Daphnia magna. *Environ Toxicol Chem* 2009, 28, 2142–2149.

77. Imhof HK, Ivleva NP, Schmid J, Niessner R, and Laforsch C. Contamination of beach sediments of a subalpine lake with microplastic particles. *Curr Biol* 2013, 23, R867–R868.

78. Karami A, Romano N, Galloway T, and Hamzah H. Virgin microplastics cause toxicity and modulate the impacts of phenanthrene on biomarker responses in African catfish (Clarias gariepinus). *Environ Res* 2016, 151, 58–70.

79. Fenchel T. Suspension feeding in ciliated protozoa: Feeding rates and their ecological significance. *Microb Ecol* 1980, 6, 13–25.

80. Rochman CM, Hoh E, Kurobe T, and Teh SJ. Ingested plastic transfers hazardous chemicals to fish and induces hepatic stress. *Sci Rep* 2013, 3, 3263.

81. Lithner D, Damberg J, Dave G, and Larsson A. Leachates from plastic consumer products –screening for toxicity with Daphnia magna. *Chemosphere* 2009, 74(9), 1195–1200.

82. Besseling E, Wegner A, Foekema EM, van den Heuvel-Greve MJ, and Koelmans AA. Effects of microplastic on fitness and PCB bioaccumulation by the Lugworm Arenicola marina (L.). *Environ Sci Technol* 2013, 47(1):593–600.

83. Paul-Pont I, Lacroix C, González Fernández C, Hégaret H, Lambert C, Le Goïc N, Frère L et al. Exposure of marine mussels Mytilus spp. to polystyrene microplastics: Toxicity and influence on fluoranthene bioaccumulation. *Environ Pollut* 2016, 216, 724–737.

84. Gouin T, Roche N, Lohmann R, and Hodges G. A thermodynamic approach for assessing the environmental exposure of chemicals absorbed to microplastic. *Environ Sci Technol* 2011, 45, 1466–1472.

85. Koelmans AA, Bakir A, Burton GA, and Janssen CR. Microplastic as a vector for chemicals in the aquatic environment: Critical review andmodel-supported reinterpretation of empirical studies. *Environ Sci Technol* 2016, 50, 3315–3326.

86. Besseling E, Wang B, Lurling M, and Koelmans AA. Nanoplastic affects growth of S. obliquus and reproduction of D. magna. *Environ Sci Technol* 2014, 48, 12336–12343.

87. Pohnert G, Steinke M, and Tollrian R. Chemical cues, defence metabolites and the shaping of pelagic interspecific interactions. *Trends Ecol Evol* 2007, 22, 198–204.

88. Lürling M and Scheffer M. Info-disruption: Pollution and the transfer of chemical information between organisms. *Trends Ecol Evol* 2007, 22, 374–379.

89. Kirstein IV, Kirmizi S, Wichels A, Garin-Fernandez A, Erler R, Löder M, Gerdts G. Dangerous hitchhikers? Evidence for potentially pathogenic Vibrio spp. on microplastic particles. *Mar Environ Res* 2016, 120, 1–8.

90. Oberbeckmann S, L€oder MGJ, and Labrenz M. Marine microplastic-associated biofilms – a review. *Environ Chem* 2015, 12, 551–3562.

91. Rist S, and Hartmann NB. *Aquatic Ecotoxicity of Microplastics and Nanoplastics: Lessons Learned from Engineered Nanomaterials, The Handbook of Environmental Chemistry*, vol 58, 25–50.

92. Wiesner MR, Lowry GV, Casman E, Bertsch PM, Matson CW, Di Giulio RT, Liu J, and Hochella MF. Meditations on the ubiquity and mutability of nano-sized materials in the environment. *ACS Nano* 2011, 5, 8466–8470.

93. Phuong NN, Zalouk-Vergnoux A, Poirier L, Kamari A, Chatel A, Mouneyrac C, and Lagarde F. Is there any consistency between the microplastics found in the field and those used in laboratory experiments? *Environ Pollut* 2016, 211, 111–123.

94. Wickson F, Hartmann NB, Hjorth R, Hansen SF, Wynne B, and Baun A. Balancing scientific tensions. *Nat Nanotechnol* 2014, 9, 870–870.

95. Lin D, Tian X, Wu F, and Xing B. Fate and transport of engineered nanomaterials in the environment. *J Environ Qual* 2010, 39, 1896.

96. Desai MP, Labhasetwar V, Walter E, Levy RJ, and Amidon GL. The mechanism of uptake of biodegradable microparticles in Caco-2 cells is size dependent. *Pharm Res* 1997, 14(11), 1568–1573.

97. Sørensen SN, and Baun A. Controlling silver nanoparticle exposure in algal toxicity testing –a matter of timing. *Nanotoxicology* 2015, 9, 201–209.

98. Dris R, Gasperi J, and Tassin B. Sources and fate of microplastics in urban areas: A focus on Paris Megacity. In: Wagner M, and Lambert S (eds) *Freshwater Microplastics: Emerging Environmental Contaminants?* Springer, Heidelberg, 2017.

99. Verschoor AJ. *Towards a definition of microplastics. Considerations for the specification of physico-chemical properties.* RIVM Letter Report 2015–0116, 2015.

100. Miklos D, Obermaier N, and Jekel M. Mikroplastik: Entwicklung eines Umweltbewertungskonzepts – ErsteÜ berlegungen zur Relevanz von synthetischen Polymeren in der Umwelt. *UBA Texte* 32/2016. ISSN 2016, 1862–4804.

101. Eriksen M, Thiel M, Prindiville M, and Kiessling T. Microplastic: What are the solutions? In: Wagner M, and Lambert S (eds) *Freshwater Microplastics: Emerging Environmental Contaminants?* Springer, Heidelberg, 2017.

102. Oehlmann J, Schulte-Oehlman U, Kloas W, Jagnytsch O, Lutz I, Kusk KO, Wollenberger L et al. A critical analysis of the biological impacts of plasticizers on wildlife. *Phil Trans R Soc B* 2009, 364(1526), 2047–2062.

103. Nizzetto L, Bussi G, Futter MN, Butterfield D, and Whitehead PG. A theoretical assessment of microplastic transport in river catchments and their retention by soils and river sediments. *Environ Sci Processes Impacts* 2016,18, 1050–1059.

104. Milliman JD, Huang-ting S, Zuo-sheng Y, and Mead RH. Transport and deposition of river sediment in the Changjiang estuary and adjacent continental shelf. *Cont Shelf Res* 1985, 4(1–2), 37–45.

105. Naden PS, Murphy JF, Old GH, Newman J, Scarlett P, Harman M, Duerdoth CP et al. Understanding the controls on deposited fine sediment in the streams of agricultural catchments. *Sci Total Environ* 2016, 547, 366–381.

106. Chakrapani GJ. Factors controlling variations in river sediment loads. *Curr Sci* 2005, 88(4), 569–575.

107. Williams AT, and Simmons SL. Sources of Riverine Litter: The River Taff, South Wales, UK. *Water Air and Soil Pollut*, 1999, 112(1–2), 197–216.

108. Galgani F, Fleet D, Van Franeker J, Katsanevakis S, Maes T, Mouat J, Oosterbaan L et al. *Marine Strategy Framework Directive Task Group 10 Report Marine litter* April 2010.

109. Cole M, Lindeque P, Halsband C, and Galloway TS. Microplastics as contaminants in the marine environment: A review. *Mar Pollut Bull* 2011, 62(12), 2588–2597.

110. Corcoran PL, Biesinger MC, and Grifi M. Plastics and beaches: A degrading relationship. *Mar Pollut Bull* 2009, 58(1), 80–84.

111. Teuten EL, Rowland SJ, Galloway TS, and Thompson RC. Potential for Plastics to Transport Hydrophobic Contaminants. *Environ Sci Technol* 2007, 41(22), 7759–7764.

112. Teuten EL, Saguing JM, Knappe DRU, Barlaz MA, Jonsson S, Björn A, Rowland SJ et al. Transport and release of chemicals from plastics to the environment and to wildlife. *Phil Trans R Soc B* 2009, 364(1526), 2027–2045.

113. Reisser J, Shaw J, Hallegraeff G, Proietti M, Barnes DKA, Thums M, Wilcox C et al. Millimeter-Sized Marine Plastics: A New Pelagic Habitat for Microorganisms and Invertebrates. *PLOS ONE* 2014, 9(6), e100289. doi:10.1371/journal.pone.0100289

114. Carpenter EJ, and Smith Jr. KL. Plastics on the Sargasso Sea Surface. *Science* 1972, 175(4027), 1240–1241.

115. De Witte B, Devriese L, Bekaert K, Hoffman S, Vandermeersch G, Cooreman K, and Robbens J. Quality assessment of the blue mussel (*Mytilus edulis*): Comparison between commercial and wild types. *Mar Pollut Bull* 2014, 85(1), 146–155.

116. Van Cauwenberghe L, and Janssen CR. Microplastics in bivalves cultured for human consumption. *Environ Pollut* 2014, 193, 65–70.

117. Wright SL, Rowe D, Thompson RC, and Galloway TS. Microplastic ingestion decreases energy reserves in marine worms. *Curr Biol* 2013, 23(23), R1031–R1033.
118. Jeong C-B, Kang H-M, Lee M-C, Kim D-H, Han J, Hwang D-S, Soussi S et al. Adverse effects of microplastics and oxidative stress-induced MAPK/Nrf2 pathway-mediated defense mechanisms in the marine copepod Paracyclopina nana. *Sci Rep* 2017, 7, 41323.
119. Bakir A, Rowland SJ, and Thompson RC. Enhanced desorption of persistent organic pollutants from microplastics under simulated physiological conditions. *Environ Pollut* 2014, 185, 16–23.

CHAPTER 12

Microplastics in Freshwater Systems
A Review on Its Accumulation and Effects on Fishes[*]

Asif Raza

CONTENTS

12.1 INTRODUCTION

Plastic materials are of vital use, being noncorrosive, durable, nonreactive, lightweight and easy to handle, and its cheap manufacturing cost has made it a material of choice. Plastic production continues to accelerate and the reason behind this is the adoption of a use-and-dispose culture by almost all the developed and developing countries. Annual plastic production has increased from 1.5 million tonnes in the 1950s to 288 million tonnes in 2012 [1], with only 9% of plastics being currently recycled in the US [2]. The nonrecycled plastic is being disposed of in dump yards; a major proportion of it is thrown as debris into water bodies. including oceans and rivers. It is estimated that 275 million metric tonnes of plastic waste is being generated each year (based on reports from 192 coastal countries in the year 2010). Due to a variety of physical, chemical and biological factors, these nonrecycled plastics in the water bodies break down to form microplastics (MPs). MPs from personal-care products are one of the potential sources of direct addition to freshwater streams. Most of the studies have occurred in marine water systems, but little data are available on the abundance and distribution of MPs in freshwater systems;

---

[*] Previously published as Open Access article in Preprints 2018, 2018100696 (doi: 10.20944/preprints201810.0696.v1).

however, MP pollution is found in estuarine water and freshwater systems [3,4]. Most studied impacts of plastic debris on biota are their physical effects such as entanglement, ingestion and suffocation/asphyxia [5–7]. These microplastics are often consumed by fishes via a variety of methods and cause adverse effects leading to mortality, neurotoxicity, cytotoxicity, liver stress, behavioral changes, oxidative stress, genotoxicity, etc. [80]. Plastic abundance was found within the stomach, gut and intestines of the fishes. The objective of this chapter is to review the current knowledge of MP contamination in freshwater and its effects on fishes. A summary of its occurrence and distribution is also discussed, along with explored knowledge of its effects on fish health have been presented in this study. Several challenges have been discussed, and suggestions are provided for further research work.

12.2 MICROPLASTICS OVERVIEW: TYPES AND SOURCES

At first, the term "microplastics" was used for the plastic matters in the range of 20 μm [8]. But later, this range was widened in the range of smaller than 5 mm [9] with the upper limit of 1 mm (1000 μm), stated by [10]. However, microplastics (MPs) are commonly defined as plastic particles having the size less than 5 mm [11–13]. This chapter concerns primarily the presence of MPs in freshwater bodies and its impacts on fishes. Research efforts on the accumulation and impacts in the freshwater system are very much less than the marine and terrestrial systems [14,15]. The concentration of MPs is constantly increasing in the aquatic environment owing to a tremendous increase in the production of plastics, with a total global production of 335 million tonnes in 2016 [16]. Most of the authors have concluded that the primary sources of MPs are effluents from wastewater treatment plants (WWTPs), sewage sludge, shipping activities, atmospheric fallouts, direct disposal from the public, beach littering and runoffs from agricultural, recreational and urban areas (Figure 12.1).

Although the data are so far unavailable, the runoffs from industrial plastic production sites can be taken as an additional source. The products such as facial scrubs have been identified as a potential source of MPs in water bodies. A study shows that the size range

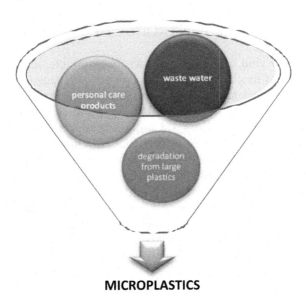

MICROPLASTICS

FIGURE 12.1   Origin of microplastics.

of four personal-care and cosmetic product wastes were in the range of 63–125 μm, 125–250 μm, 250–500 μm and 500–2000 μm [17]. Generally, MPs are classified as primary or secondary on the basis of their production.

Primary MPs are the ones having size <5 mm and mainly are originated from textiles, medicines, toothpaste and variety of other personal-care products like facial products and scrubs [17,18]. The range of primary MPs and their types mainly consists of fragments [19], fibers [19], films and foams [20]. Secondary MPs can be originated by the fragmentation of big plastic materials degradation. They are derived from the degradation of larger plastic debris through mechanical forces, thermal degradation, photolysis, thermo-oxidation and biodegradation processes [3], for example, synthetic fibers from washing clothes [21]. Secondary MPs arising by washing clothes are generally polyester, acrylic, and polyamide, which can be more than 100 fibers/L of effluent [21,22] (Figures 12.2 and 12.3).

12.3 MICROPLASTICS

However, we also describe plastics according to their basic chemical structure as polyethylene (PE), polypropylene (PP), polyamide (PA), polyvinyl chloride (PVC), polystyrene (PS), polyurethane (PU) and polyethylene terephthalate (PET) [10,13]. These are the structures extensively found in the majority of research.

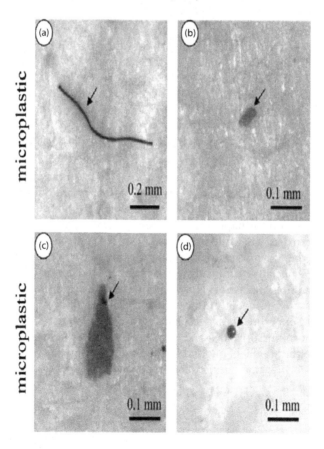

FIGURE 12.2   Photographs of microplastics from fish.

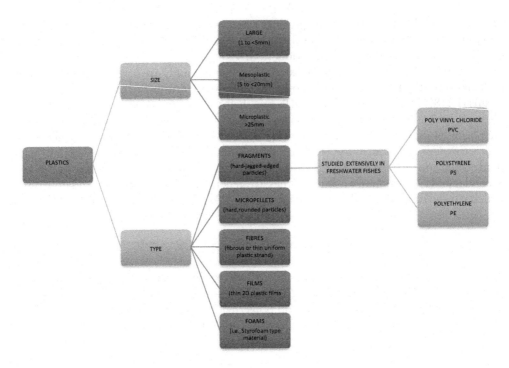

FIGURE 12.3   Flow-diagram of types of plastics information collected (Adapted from Anderson PJ et al. Environ Pollut 2017, 225, 223–231; Lee KW et al. Environ Sci Technol 2013, 47(19), 11278–11283; Hidalgo-Ruz V. and Thiel M. In M. Bergmann, L. Gutow, M. Klages (eds.), Marine Anthropogenic Litter, Heidelberg, New York, Dordrecht, London: Springer Cham, 2015, pp. 429–447.)

12.4 METHODS OF INGESTION OF MPS BY FISHES

Ingestion is one of the most common factors associated with plastic debris, having been reported in more than 270 taxa [24] from a variety of trophic levels [18]. One of the most affected taxa is fish. Plastic and other debris may be intentionally ingested by fish [18,24]. Incidental ingestion happens with the ingestion of natural food items [25], or through trophic transfer, when the fish consumes prey that has already ingested plastic debris [26,27]. On the other hand, intentional ingestion occurs when the plastic material is mistaken for food, especially bottom algae and fragment-like foods [28]. Evidence suggests that intentional ingestion of plastic is most common in fish. For instance, marks left in large plastic debris suggest fish frequently attack and bite plastic items present in the environment [29], and laboratory experiments suggest fish larvae feed preferentially on plastic particles when exposed to both microplastics and natural food [30]. The ingested MPs usually get accumulated inside the stomach, gut and intestinal lining of the fishes, which we examine to observe whether or not the ingestion has happened (Table 12.1; Figure 12.4).

12.4.1 Data Analysis

Plastic debris ingestion in fish from different freshwater habitats like rivers, estuaries and lakes from different locations across the globe was examined by different researchers, and

TABLE 12.1 Plastic Debris Ingestion in Fish from Freshwater Habitats (Rivers, Estuaries and Lakes)

Environment	Location	Species	Frequency %	Reference
Lake	Lake Victoria (Africa)	*Lates niloticus*	20	[31]
Lake	Lake Victoria (Africa)	*Oreochromis niloticus*	20	[31]
Estuary	Goina Estuary (Brazil)	*Cathorops spixi*	18	[32]
Estuary	Goina Estuary (Brazil)	*Cathorops agassizzi*	33	[32]
Estuary	Goina Estuary (Brazil)	*Sciades herzbergii*	18	[32]
Estuary	Goina Estuary (Brazil)	*Stellifer brasiliensis*	6.9	[33]
Estuary	Goina Estuary (Brazil)	*Stellifer stellifer*	9.2	[33]
Estuary	Goina Estuary (Brazil)	*Eugeress brasilianus*	16.3	[34]
Estuary	Goina Estuary (Brazil)	*Eucinostomos melanopterus*	9.2	[34]
Estuary	Goina Estuary (Brazil)	*Diapterus rhombeus*	11.4	[34]
River	Brazos River Basin (US)	*Leponis megalotis*	44	[25]
River	Brazos River Basin (US)	*Lepomis macrochirus*	45	[25]
River	7 Rivers (France)	*Gobio gobio*	9.5	[35]
			4.2	
River/ lake	Various (USA)	26 species	5–29	[36]
River	Pajeu River (Brazil)	*Haplosternum littorale*	83	[37]

Source: Silva-Cavalcanti J et al. *Environ Pollut* 2017, 221, 218–226.

Note: The frequency indicates the percentage of individuals observed with the plastic debris inside the gut.

it was found that a lot of species ingest MPs. The data provided in Table 12.1 show the percentage of individuals observed with plastic debris inside the gut, shown as frequency percentage. Among the estuary species, the frequency percentage was highest in *Cathorops agassizzi* collected from Goina estuary (Brazil) in the year 2011; it had ingested debris percentage of approximately 33%, which is quite high. Other species like *Cathorops spixi* and *Sciades herbergii* showed the frequency equal to 18%. The genus *Stellifer* had ingested debris frequency percentage between 6% to 9%; specifically, *Stellifer brasilliensis* (found lowest among estuary species observed) and *Stellifer stellifer* had 6.9% and 9.2%, respectively, of ingested debris frequency. Other species like *Eugeress brasilianus, Eucinostomos melanopterus,* and *Diapterus rhombeus* were found with 16.3%, 9. 2% and 11.4% of frequency debris, respectively.

The reports from the species thriving in Lake Victoria on the African continent showed an almost static frequency percentage of 20% among the studied species *Lates niloticus* and *Oreochromis niloticus,* studied quite recently in the year 2016 [31]. There was a high frequency of ingested debris among the species of river habitats; so far, *Haplosternum littorale* collected from the Pajeu River (Brazil) had a remarkably high (and the highest) frequency percentage of 83%. It is predicted to be hazardous for humans via the food chain; however, it is just a hypothesis as actual effects are yet to be analyzed. Other species collected from the Brazos River Basin (US) also showed high percentages of

FIGURE 12.4 Showing some of the studied species having a considerable frequency of ingested MPs (e.g., *Gobio gobio, Sciades herzbergii, Cathorops spixii, Diapterus rhombeus, Stellifer stellifer, Cathorops agassizii* and *Lates niloticus*).

ingested debris frequencies: *Leponis megalotis* and *Lepomis macrochirus* were observed with 44% and 45% frequencies, respectively [25]. Sanchez examined *Gobio gobio* from seven rivers of France in the year 2014 and found a range of frequency percentage of 9.5%–4.2% [35]. The overall result shows that the species thriving in the rivers are mostly affected by the MPs contamination. The reason behind this could be because the river is vulnerable to various sewage discharges along with factory wastes, so the chance of contamination is highest.

According to the hypothesis, the biomagnification of MPs is likely to be highest through river water species, and the consumption of the infected fish with MPs can be hazardous for humans also.

12.5 EFFECTS OF MICROPLASTICS ON FISHES

The effects of MP contamination on fish health are not yet fully understood. The ingestion of MPs by fishes can get accumulated in their digestive tract, which can cause starvation

TABLE 12.2   Summary of MP Effects

Effect	Description	References
Increased reactive oxygen species (ROS)	Ingested microplastics have shown to increase free radicals, which leads to cellular and DNA damage.	[44]
Reduced feeding or filtering	Animals containing microplastics in their digestive tracts were found to eat less, resulting in lower energy levels and fat reserves.	[42,43]
Immune response	Microplastics in animal tissue can induce an immune response leading to inflammation.	[45,46]
Hepatic damage	Owing to metabolic stress caused by microplastics, as well as pollutants accumulating on the surface, liver damage has been found in some organisms.	[47,48]
Reduced gamete quality	Lower gamete quality causes fewer offspring to be produced and decreased fecundity.	[49]
Mortality	Owing to a combination of the physical and physiological effects of microplastic particles on certain individuals, fatality is increased.	[23]

Source: Bouwman H et al. *Microplastics in Freshwater Environments*, 1 edn., Water Research Commission, Republic of South Africa, 2018, pp. 5–22.

because of the false sensation of satiation or even perforation of the gastrointestinal tract. It may also pass to predators including humans [38–40]. Internal and digestive enzyme systems may get damaged; even reproduction can be affected because of MPs digestion [41–43]. Examples of studies are listed in Table 12.2.

As MPs act as a sponge and provide surface area for various bio-organic or inorganic toxic substances; the ingestion of these adsorbed toxin-containing MPs could be a serious health issue for the fishes. The negative effect of toxins on fish health was demonstrated by [47,48]. Tiny particles of low-density polyethylene (LDPE) were exposed to environmental bay conditions for 3 consecutive months and then fed to fishes. After 2 months, the tissues of fish had a greater concentration of PBTs and showed signs of liver stress, glycogen depletion, fatty vacuolation and cell necrosis [47,48] (Figure 12.5).

A total of 21 studies reporting ecotoxicological effects of MPs were identified. Fishes may ingest MPs either directly or by the prey containing these particles [51]. Overall documented effects of MPs on fishes include reduction of feeding activity [52,53], oxidative stress [54], genotoxicity [54], neurotoxicity [55–58], growth delay [54,59,60], reduction in reproductive fitness [23,61] and ultimately death [23,59,61,62]. The representation of ecotoxicological effects is shown in Figure 12.6.

12.6 RECENT GLOBAL ACTIONS ON MPs

- In the year 2012, Unilever decided to wipe out microplastics from all of its personal-care products by the year 2015.
- The report of United Nations Environment Program (UNEP) in the year 2014 recommended to increase the efforts to understand the effects of MPs, its capacity to absorb and transfer persistent, toxic and bioaccumulating chemicals.

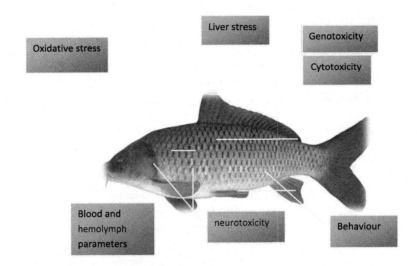

FIGURE 12.5   Some of the effects of MPs on freshwater fishes.

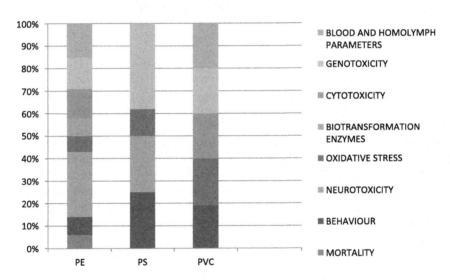

FIGURE 12.6   Ecotoxicological effects of MPs on fishes based on 21 studies. Studies were defined according to types of MPs and their effects. (Data from de Sá LC et al. *Science of The Total Environment* 645, 2018, 1029–1039.)

- The states of Illinois and California (US) passed bills to impose bans on manufacturing and selling of personal-care products with microplastic beads, in the year 2014 and 2015, respectively.
- The US has banned the production of personal-care products containing MPs in the year 2017 and also has decided to stop the sale of drugs containing MPs.
- In the year 2018, the United Kingdom imposed a ban on the manufacture of personal-care products containing microplastic beads.
- From the year 2020, countries including Sweden, Finland, France, Iceland, Ireland, Luxembourg and Norway will ban the sale of cosmetics with MPs and have called on the EU for an EU-wide ban.

- WHO, in the March 15, 2018, edition of the newspaper article of *The Guardian* reported that WHO is considering launching a health review in response to a study where MPs were found in more than 90% of some of the popular packaged water brands.

12.7 CONCLUSIONS

This chapter compiled the comprehensive information about the importance of the study of MP contamination in freshwater along with the ways of accumulation and effects on fishes. In this regard, the following topics were discussed: (a) microplastics overview, types and sources, (b) methods of ingestion of MPs by fishes and (c) effects of MPs on fishes.

From the literature, it can be concluded that MPs are a dormant hazard for aquatic organisms and their quantity is increasing day by day; it is the reason behind the several abnormalities in the behavior and health of fishes. Fishes ingest MPs intentionally and sometimes unintentionally, and the MPs get accumulated in the gastrointestinal tract and stomach of biota. PE, PS and, furthermore, PVC are among the most extensively studied MPs obtained from inside the freshwater fishes. This chapter shows the data on the ingestion of MPs by fishes of different freshwater bodies and also provides graphs to show the effects of these accumulated particles on fishes.

Based on the investigation, the following conclusions can be drawn:

- More attention is needed toward freshwater MP studies.
- Rules are needed to counter the generation of MPs in water bodies.
- There should be a ban or monitoring over the production of personal-care products containing MPs, as they are one of the primary sources.
- Toxic effects and biomagnifications of MPs through food chains need to be evaluated comprehensively.
- Better understandings of MPs effects on humans is needed.
- Techniques are needed to filter out MPs from wastewater in treatment plants.
- Researchers should establish techniques of detection and filtration of MPs from water at a satisfactory level.
- Further studies should be directed toward prevention, awareness and reduction and counter methods.

REFERENCES

1. Plastics Europe. Plastics—the Facts 2013: an Analysis of European Latest Plastics, 2013.
2. EPA. 2014. Plastics Common Wastes & Materials US EPA (WWW Document).
3. Zhao S, Zhu L, and Li, D. Microplastic in three urban estuaries, China. *Environ Pollut* 2015, 206, 597–560.
4. Su L, Xue Y, Li L, Yang D, Kolandhasamy P, Li D, and Shi H. Microplastics in Taihu Lake, *China. Environmetnal pollution* 2016, 216. 711–719.
5. Barnes DKA, Galgani F, Thompson RC, and Barlaz M. Accumulation and fragmentation of plastic debris in global environments. *Philos Trans R Soc B-Biological Sci* 2009, 364, 1985–1998.

6. Ryan PG, Moore CJ, van Franeker JA, and Moloney CL. Monitoring the abundance of plastic debris in the marine environment. *Philos Trans R Soc Lond B Biol Sci* 2009, 364, 1999e–2012. Sci. Process. Impacts 18, 788e795. http://dx.doi.org/10.1039/C6EM00234J.

7. Sigler M. The effects of plastic pollution on aquatic wildlife: Current situations and future solutions. *Water Air Soil Pollut* 2014, 225, 2184. http://dx.doi.org/

8. Thompson RC, Olsen Y, Mitchell RP, Davis A, Rowland SJ, John AWG, McDonigle D, and Rusell AE. Lost at sea: Where is all the plastic? *Science* 2004, 304, 838. 220–224.

9. Arthur C, Baker J, and Bamford H. Proceedings of the International Research Workshop on the Occurrence, *Effects and Fate of Microplastic Marine Debris.* In: NOAA Technical memorandum NOS-OR&R-30, 2009, p. 49.

10. Van Cauwenberghe L, Devriese L, Galgani F, Robbens J, and Janssen CR. Microplastics in sediments: A review of techniques, occurrence and effects. *Mar Environ Res* 2015, 111, 5–17.

11. Betts K. Why small plastic particles may pose a big problem in the oceans. *Environ Sci Technol* 2008, 42(24), 8995–8995.

12. Fendall LS, and Sewell MA. Contributing to Marine Pollution by Washing YourFace:Microplastics in Facial Cleansers. *Mar Pollut Bull* August 2009, 58(8), 1225–1228. doi:10.1016/j.marpolbul.2009.04.025

13. Hidalgo-Ruz V, Gutow L, Thompson RC, Thiel M. Microplastics in the marineenvironment: A review of the methods used for identification and quantification. *Environ Sci Technol* 2012, 46(6), 3060–3075.

14. Thompson RC, Moore CJ, VomSaal FS, and Swan SH. Plastics, the environment andhuman health: Current consensus and future trends. *Philos Trans R Soc Lond Ser B Biol Sci* 2009, 364, 2153–2166.

15. Wagner M, Scherer C, Alvarez-Mu~noz D, Brennholt N, Bourrain X, Buchinger S, Fries E et al. Microplastics infreshwater ecosystems: What we know and what we need to know. *Environ Sci Eur* 2014, 26, 12. http://dx.doi.org/10.1186/s12302-014-0012-7

16. Plastics Europe. Plastics – The Facts 2017. Plastics Europe, Brussels, 2017, pp. 44.

17. Browne MA. In: Bergmann M, Gutow L, Klages M (Eds.). *Marine Anthropogenic Litter.* Springer International Publishing, Cham, 2015, pp. 229e–244.

18. Cole M, Lindeque P, Halsband C, and Galloway TS. Microplastics as contaminantsin the marine environment: A review. *Mar Pollut Bull* 2011, 62(12), 2588e–2597.

19. Rummel CD, Adolfsson-Erici M, Jahnke A, and MacLeod M. No measurable "cleaning" of polychlorinated biphenyls from Rainbow Trout in a 9 week depura-tionstudy with dietary exposure to 40% polyethylene microspheres. *Environ* 2016, 18,788–795.

20. Anderson PJ, Warrack S, Langen V. Challis JK, Hanson ML, and Rennie MD. Microplastic contamination inLake Winnipeg, Canada. *Environ Pollut* 2017, 225, 223–231. Doi: 10.1007/s11270-014-2184-6

21. Browne MA, Crump P, Niven SJ, Teuten E, Tonkin A, Galloway T, Thompson R. Accumulation of microplastic on shorelines worldwide: Sources and sinks. *Environ Sci Technol* 2011, 45, 9175–9179.

22. Habib D, Locke DC, Cannone LJ. Synthetic fibers as indicators of municipal sewage sludge,sludge products, and sewage treatment plant effluents. *Water Air Soil Pollut* 1998, 103, 1–8.

23. Lee KW, Shim WJ, Kwon OY, and Kang JH. Size-dependent effects of micro polystyrene particles in the marine copepod Tigriopus japonicas. *Environ Sci Technol* 2013, 47(19), 11278–11283.

24. Laist DW. *Impacts of Marine Debris: Entanglement of Marine Life in Marine Debris Including a Comprehensive List of Species with Entanglement and Ingestion Records, in: Marine Debris.* Springer, 1997, p. 99e–139

25. Peters CA, and Bratton SP. Urbanization is a major influence on microplastic ingestion by sunfish in the Brazos River Basin, Central Texas, USA. *Environ Pollut* 2016, 210, 380e–387. http://dx.doi.org/10.1016/j.envpol.2016.01.018

26. Cedervall T, Hansson L-A, Lard M, Frohm B, and Linse S. Food chain transport of nanoparticles affects behaviour and fat metabolism in fish. *PLOS ONE* 2012, 7, e32254. http://dx.doi.org/10.1371/journal.pone.0032254.

27. Mattsson K, Ekvall MT, Hansson L-A, Linse S, Malmendal A, and Cedervall T. Altered behavior, physiology, and metabolism in fish exposed to polystyrenenanoparticles. *Environ Sci Technol* 2015, 49, 553e–561. http://dx.doi.org/10.1021/es5053655

28. Ivar do Sul JA, Costa MF. Marine debris review for Latin America and the wider Caribbean region: From the 1970s until now, and where do we go from here? *Mar Pollut Bull* 2007, 54, 1087e–1104.

29. Carson HS. The incidence of plastic ingestion by fishes: From the prey's perspective. *Mar Pollut Bull* 2013, 74, 170e–174.

30. L€onnstedt OM, Ekl€ov P. Environmentally relevant concentrations of microplastic particles influence larval fish ecology. *Science* 2016, 352, 1213e–1216. http://dx.doi.org/10.1126/science.aad8828

31. Biginagwa FJ, Mayoma BS, Shashoua Y, Syberg K, and Khan FR. First evidence of microplastics in the African great lakes: Recovery from lake Victoria nile perch and nile tilapia. *J Gt Lakes Res* 2016, 42, 146e–149. http://dx.doi.org/10.1016/j.jglr.2015.10.012

32. Possatto FE, Barletta M, Costa MF, Ivar do Sul JA, and Dantas DV. Plasticdebris ingestion by marine catfish: An unexpected fisheries impact. *Mar Pollut Bull* 2011, 62, 1098e–1102. http://dx.doi.org/10.1016/j.marpolbul.2011.01.036. Production, Demand and Waste Data, Octubre 2013.

33. Dantas DV, Barletta M, and Da Costa MF. The seasonal and spatial patterns of ingestion of polyfilament nylon fragments by estuarine drums (Sciaenidae). *Environ Sci Pollut Res* 2012, 19, 600e–606.

34. Ramos JA, Barletta M, and Costa MF. Ingestion of nylon threads by Gerreidae while using a tropical estuary as foraging grounds. *Aquat Biol* 2012, 17, 29e–34.

35. Sanchez W, Bender C, and Porcher J-M. Wild gudgeons (Gobio gobio) from French rivers are contaminated by microplastics: Preliminary study and first evidence. *Environ Res* 2014, 128, 98e–100. http://dx.doi.org/10.1016/j.envres.2013.11.004

36. Phillips MB, and Bonner TH. Occurrence and amount of microplastic ingested by fishes in watersheds of the Gulf of Mexico.*Mar Pollut Bull* 2015, 100, 264e–269. http://dx.doi.org/10.1016/j.marpolul.2015.08.041

37. Silva-Cavalcanti JS, Silva JDB, de França EJ, de Araújo MCB, Gusmão F. Microplastics ingestion by a common tropical freshwater fishing resource. *Environ Pollut* 221, 2017, 218–226, ISSN 0269–7491, https://doi.org/10.1016/j.envpol.2016.11.068

38. Farrell P, and Nelson K. Trophic level transfer ofmicroplastic: Mytilus edulis (L.) to Carcinus maenas (L.). *Environ Pollut* 2013, 177, 1–3.

39. Seltenrich N. New link in the food chain? Marine plastic pollution and seafood safety. *Environ Health Perspect* 2015, 123, A35–A41.

40. Sharma S, and Chatterjee S. Microplastic pollution, a threat to marine ecosystem and human health: A short review. *Environl Sci Pollut Res* 2017, 24, 21530–21547.

41. Talvitie J, Heinonen M, Pääkkönen J-P, Vathera E, Setalälä O, and Vahala R. Do wastewater treatment plants act as a potential point source of microplastics? Preliminary study in the coastal Gulf of Finland. *Baltic Sea Water Sci Technol* 2015, 72, 1495–1504.

42. Wright SL, Rowe D, Thompson RC, and Galloway TS. Microplastic ingestion decreases energy reserves in marine worms. *Curr Biol* 2013, 23(23), R1031–R1033.

43. Wright SL, Thompson RC, and Galloway TS. The physical impacts of microplastics on marine organisms: A review. *Environ Pollut* 2013, 178, 483–492.

44. Bhattacharya P, Turner JP, and Ke PC. Physical adsorption of charged plastic nanoparticles affects algal photosynthesis. *J Phys Chem* 2010, 114, 16556–16561.

45. von Moos N, Burkhardt-Holm P, and Kohler A. Uptake and effects of microplastics on cells and tissue of the blue mussel *Mytilus edulis* L. after an experimental exposure. *Environ Sci Technol* 2012, 46, 11327–11335.

46. Köhler A. Cellular fate of organic compounds in marine invertebrates. *Comp BiochemPhysiol—Part A: Mol Integr Physiol* 2010, 157(Supplement), 8–11.

47. Rochman CM. Plastics and priority pollutants: A multiple stressor in aquatic habitats. *Environ Sci Technol* 2013, 47, 2439–2440.

48. Rochman CM, Hoh E, Kurobe T, and Teh SJ. Ingested plastic transfers hazardous chemicals to fish and induces hepatic stress. *Sci Rep* 2013, 3, 3263. h ttp://dx.doi. org/10.1038/srep03263

49. Sussarellu R, Soudant P, Lambert C, Fabioux C, Corporeau C, Laot C et al. Microplastics: Effects on oyster physiology and reproduction. Platform presentation, *International workshop on fate and impact of microplastics in marine ecosystems (MICRO2014)*, 2014, 13–15 January 2014. Plouzane (France).

50. Bouwman H, Minnaar K, Bezuidenhout C, and Verster C. *Microplastics in Freshwater Environments*, 1 edn., Water Research Commission, Republic of South Africa, 2018, pp. 5–22.

51. Desforges JPW, Galbraith M, Dangerfield N, and Ross PS. Widespread distribution of microplastics in subsurface seawater in the NE Pacific Ocean. *Mar Pollut Bull* 2014, 79, 94–99.

52. Besseling B, Wegner A, Foekema EM, Heuvel-Greve MJ, and Koelmans AA. Effects of microplastic on fitness and PCB bioaccumulation by the lugworm Arenicola marina(L.). *Environ Sci Technol* 2013, 47, 593–600.

53. de Sá LC, Luís LG, and Guilhermino L. Effects of microplastics on juveniles of the common goby (Pomatoschistus microps): Confusion with prey, reduction of the predatoryperformance and efficiency, and possible influence of developmental conditions.*Environ Pollut* 2015, 196, 359–362.

54. Della Torre C, Bergami E, Salvati A, Faleri C, Cirino P, Dawson KA, and Corsi I. Accumulation and embryotoxicity of polystyrene nanoparticles at early stage of development of sea urchin embryos Paracentrotus lividus. *Environ Sci Technol* 2014, 48(20), 12302–12311.

55. Oliveira M, Ribeiro A, and Guilhermino L. Effects of short-term exposure to microplastics and pyrene on Pomatoschistus microps (Teleostei, Gobiidae). *Comp Biochem Physiol A Mol Integr Physiol* 2012, 163, S20–S20.

56. Oliveira M, Ribeiro A, Hylland K, and Guilhermino L. Single and combined effects of microplastics and pyrene on juveniles (0+ group) of the common goby Pomatoschistus microps (teleostei: Gobiidae). *Ecol Indic* 2013, 34, 641–647.

57. Luis L, Ferreira P, Fonte E, Oliveira M, and Guilhermino L. Does the presence of microplastics influence the acute toxicity of chromium(VI) to early juveniles of

the common goby (Pomatoschistus microps)? A study with juveniles from two wild estuarine populations. *Aquat Toxicol* 2015, 164, 163–174.

58. Ribeiro F, Garcia AR, Pereira BP, Fonseca M, Mestre NC, Fonseca TG, Ilharco LM, and Bebianno MJ. Microplastics effects in Scrobicularia plana. *Mar Pollut Bull* 2017, 122(1–2), 379–391.

59. Au SY, Bruce TF, Bridges WC, and Klaine SJ. Responses of Hyalella Azteca to acute and chronic microplastic exposure. *Environ Toxicol Chem* 2015, 34, 2564–2572.

60. Plastic Europe. Redondo-Hasselerharm PE, Falahudin D, Peeters ETHM, and Koelmans AA. Microplastic effect thresholds for freshwater benthic macroinvertebrates. *Environ Sci Technol* 2018, 52(4), 2278–2286.

61. Cole M, Lindeque P, Fileman E, Halsband C, and Galloway TS. The impact of polystyrenemicroplastics on feeding, function and fecundity in the marine copepodCalanus helgolandicus. *Environ Sci Technol* 2015, 49, 1130–1137.

62. Mazurais D, Emande B, Quazuguel P, Severe A, Huelvan C, Madec L, Mouchel O et al. Evaluation of the impactof polyethylene microbeads ingestion in European sea bass (Dicentrarchus labrax) larvae. *Mar Environ Res* 2015, 112(Part A), 78–85.

63. de Sá LC, Oliveira M, Ribeiro F, Rocha TL, Futter MN. Studies of the effects of microplastics on aquatic organisms: What do we know and where should we focus our efforts in the future? *Sci Total Environ* 2018, 645, 1029–1039, ISSN 0048–9697, https://doi.org/10.1016/j.scitotenv.2018.07.207.

64. Browne MA, Dissanayake A, Galloway TS, Lowe DM, and Thompson RC. Ingested microscopic plastic translocates to the circulatory system of themussel, Mytilus edulis (L.). *Environ Sci Technol* 2008, 42, 5026–5031.

65. Browne MA, Galloway TS, and Thompson R. Microplastic – an emerging contaminant of potential concern? *Integr. Environ Assess Manag* 2007, 3, 559–561.

66. Browne MA, Niven SJ, Galloway TS, Rowland SJ, and Thompson RC. Microplastic moves pollutants and additives to worms, reducing functions linked to health and biodiversity. *Curr Biol* 2013, 23(23), 2388–2392.

67. Ruiz CE, Esteba MÁ, Cuestz A. Microplastics in aquatic environments and their toxicological implications for fish. In Sonia Soloneski and Marcelo L. Larramendy (Eds.)*Toxicology New Aspects to this Scientific Conundrum*, Intech Croatia, 2016, pp.113–145.

68. de Araújo F G. Microplastics ingestion by a common tropical freshwater fishing resource. *Environ Pollut* 2017, 221, 218–226, ISSN 0269–7491, https://doi.org/10.1016/j.envpol.2016.11.068.

69. Desforges JPW, Galbraith M, and Ross PS. Ingestion of microplastics by zooplankton in the Northeast Pacific Ocean. *Arch Environ Contam Toxicol* 2015, 69, 320–330.

70. Duis K, and Coors A. Microplastics in the Aquatic and Terrestrial Environment: Sources (with a Specific Focus on Personal Care Products), Fate and Effects. *Environmental Sciences Europe* January 6, 2016, 28(1), doi:10.1186/s12302-015-0069-y.

71. Li J, Liu H, Paul Chen J. Microplastics in freshwater systems: A review on occurrence, environmental effects, and methods for microplastics detection. *Water Res* 2018, 137, 362–374, ISSN 0043–1354, https://doi.org/10.1016/j.watres.2017.12.056

72. Jabeen K, Su L, Li J, Yang D, Tong C, Mu J, Shi H, Microplastics and mesoplastics in fish from coastal and fresh waters of China. *Environ Pollut* 2017, 221, 141–149, ISSN 0269–7491, https://doi.org/10.1016/j.envpol.2016.11.055

73. Lechner A, Keckeis H, Lumesberger-Loisl F, Zens B, Krusch R, Tritthart M, Glas M, and Schludermann, E. The Danube so colourful: A potpourri of plastic litter outnumbers fish larvae in Europe's second largest river. *Environ Pollut* 2014,188,177–181.

74. Lithner D, Larsson A, and Dave G. Environmental and health hazard ranking and assessment of plastic polymers based on chemical composition. *Sci Total Environ* 2011, 409, 3309–3324. http://dx.doi.org/10.1016/j.scitotenv.2011.04.038

75. Lusher AL, McHugh M, and Thompson RC. Occurrence ofmicroplastics in the gastrointestinaltract of pelagic and demersal fish from the English Channel. *Mar Pollut Bull* 2013, 67, 9499.

76. Rezania S, Park J, Md Din MF, Mat Taib S, Talaiekhozani A, Kumar Yadav K, Kamyab H. Microplastics pollution in different aquatic environments and biota: A review of recent studies. *Mar Pollut Bull* 2018, 133, 191–208, ISSN 0025–326X, https://doi.org/10.1016/j.marpolbul.2018.05.022

77. Vendel AL et al. Widespread Microplastic Ingestion by Fish Assemblages in Tropical Estuaries Subjected to Anthropogenic Pressures. *Mar Pollut Bull* 2017, 117(1–2), 448–455.

78. Wagner M, and Lambert S. *Freshwater Microplastics. Emerging Environmental Contaminants? The handbook of Environmental Chemistry* Vol 58. Springer Open, Cham, Switzerland, 2018.

79. Yu Y, Zhou D, Li Z, Zhu C. Advancement and Challenges of Microplastic Pollution in the Aquatic Environment: A Review. *Water, Air, & Soil Pollution* 2018, 2018/04/17, 140 229(5), 1573–2932. https://doi.org/10.1007/s11270-018-3788-z 10.1007/s11270-018-3788-zYu2018

80. Barboza LGA, Vethaak AD, Lavorante BRBO, Lundebye A-K, Guilhermin L. Marine microplastic debris: An emerging issue for food security, food safety and human health. *Mar Pollut Bull*, 2018, 133, 336–348.

81. Hidalgo-Ruz V., and Thiel M. The Contribution of Citizen Scientists to the Monitoring of Marine Litter. In M. Bergmann, L. Gutow, M. Klages (eds.), *Marine Anthropogenic Litter*, Heidelberg, New York, Dordrecht, London: Springer Cham, 2015, pp. 429–447.

Index

Printed in the United States
By Bookmasters